Springer Proceedings in Mathematics & Statistics

Volume 84

For further volumes:
http://www.springer.com/series/10533

Springer Proceedings in Mathematics & Statistics

This book series features volumes composed of select contributions from workshops and conferences in all areas of current research in mathematics and statistics, including OR and optimization. In addition to an overall evaluation of the interest, scientific quality, and timeliness of each proposal at the hands of the publisher, individual contributions are all refereed to the high quality standards of leading journals in the field. Thus, this series provides the research community with well-edited, authoritative reports on developments in the most exciting areas of mathematical and statistical research today.

Raffaele De Amicis · Giuseppe Conti
Editors

Future Vision and Trends on Shapes, Geometry and Algebra

Springer

Editors
Raffaele De Amicis
Giuseppe Conti
Fondazione Graphitech
Povo, Trento
Italy

ISSN 2194-1009 ISSN 2194-1017 (electronic)
ISBN 978-1-4471-6460-9 ISBN 978-1-4471-6461-6 (eBook)
DOI 10.1007/978-1-4471-6461-6
Springer London Heidelberg New York Dordrecht

Library of Congress Control Number: 2014941092

Mathematics Subject Classification: 65D17, 68U07, 68W30, 13P15, 68U10

Printed on acid-free paper

Springer is part of Springer Science+Business Media (www.springer.com)

Contents

Introduction

This volume collects the results presented in the context of the final workshop of the project SAGA—"ShApes Geometry Algebra," a four-year long Initial Training Network that had been funded by the European Commission within the Marie Curie Action. Its aim was to introduce young researchers to interdisciplinary and inter-sectorial research and development issues in the areas of classical and computational geometry, approximation theory, numerical analysis, and computer graphics and—most notably—their implications in the context of Computer Aided Design and Manufacturing (CAD/CAM) systems.

In fact, over the last few years, disciplines such as computational geometry or computer graphics have undergone a profound evolution essentially driven by the ever-increasing computational power and by the availability of new hardware and software technologies. This, in turn, is having profound effects, either directly or indirectly, on a significant range of domains including, but not limited to, digital content creation and creative industry (e.g. performing arts, film, animation movies, video gaming, digital libraries), multimodal interfaces and natural computer interaction, scientific visualisation (e.g. medicine, chemical and pharmaceuticals), policy making (e.g. planning, environment, public sector information), engineering and manufacturing (e.g. civil, manufacturing, automotive and aerospace engineering), marketing (e.g. digital signage, advertisement campaigns), finance, safety and security.

Furthermore, due to their horizontal nature, any development in fields such as computational geometry, approximation theory, numerical analysis and computer graphics brings a number of direct and intermediate implications on a very large range of which are not directly related to CAD/CAM, ranging from pedagogy to neurosciences, from digital storytelling to social and cognitive sciences, to name but a few.

The aim of the project SAGA was to bring such a broad-range perspective to the community of young scientists and, through a comprehensive four-year training programme, promote the growth of a new generation of researchers capable to address a wide range of requirements emerging from both the academic and the industrial standpoint.

A three-day long workshop, which took place in Trento, Italy at the end of 2012, marked the end of the SAGA project. The event had been articulated to give project fellows the opportunity to showcase research achievements and meet with top experts from domains of relevance, who addressed, in the context of several keynote speeches, the following research and industrial challenges:

- Detecting Hidden Curves Using Algebraic Schemes (Prof. Lorenzo Robbiano from the University of Genoa, Italy).
- Virtual Conceptual Design (Mr. Gino Brunetti from CASED—Center for Advanced Security Research Darmstadt, Germany).
- Polynomial Splines over Locally Refined Box-Partitions (Dr. Tor Dokken from SINTEF, Norway and SAGA coordinator).
- Numerical Algebraic Geometry and Differential Equations (Prof. Andrew Sommese from the University of Notre Dame, USA).
- The Case for Alternatives to Interactive Design (Dr. Thomas A. Grandine from BOEING Research and Technology, USA).

The majority of the works collected within this volume were presented during the aforementioned workshop and span across several relevant domains, including:

- Mathematical algorithms for geometric and solid modelling.
- Scientific issues in the context of classical algebraic geometry.
- Industrial applications of mathematical models, for instance required deforming surfaces when animating virtual characters, to automatically compare images of handwritten signatures, or to improve control of NC machines.

Their multifaceted nature well reflects the cross-disciplinary nature of SAGA, which was funded upon four main pillars: change of representation, geometric computing and algebraic tools, algebraic geometry for CAD applications and practical industrial problems.

Raffaele De Amicis
Giuseppe Conti

Future Shape Modelling Scenarios

Current Trends and Possible Evolutions Within the Industrial Domains of Engineering and Manufacturing

Raffaele De Amicis, Giuseppe Conti and André Stork

Abstract The purpose of this chapter is to present a broad analysis of the recent developments that characterize the domain of CAD/CAM/CAE and, starting from the current state of the art in terms of scientific achievements, to analyse technological trends that regard the interactive visualization domain, and eventually define a number of possible scenarios that are likely to unfold in the next few years in the context of 3D shape modelling for industrial applications.

Keywords CAx · 3D modelling · Virtual engineering · Future trends

1 Introduction

Global competition is continuously forcing manufacturers of industrial products within the automotive, aerospace or other production engineering domains, to introduce faster, cheaper and more agile processes to facilitate the creation of better performing and more personalized products. This trend has produced a significant strain to the whole production development process. In recent years the different models used to represent the various facets of the design and manufacturing processes have been profoundly transformed by a continuous evolving technological landscape.

R. De Amicis(✉) · G. Conti
Fondazione Graphitech, Via Alla Cascata 56C, 38123 Trento, Italy
e-mail: raffaele.de.amicis@graphitech.it

G. Conti
e-mail: gc@giuseppeconti.com

A. Stork
Fraunhofer Institute for Computer Graphics Research IGD, Fraunhoferstr. 5,
64283 Darmstadt, Germany
e-mail: andre.stork@igd.fraunhofer.de

R. De Amicis and G. Conti (eds.), *Future Vision and Trends on Shapes, Geometry and Algebra*, Springer Proceedings in Mathematics & Statistics 84,
DOI: 10.1007/978-1-4471-6461-6_1, © Springer-Verlag London 2014

Over the past few years, the domains of Computer Aided Technologies (CAx)—in general—and Computer Aided Manufacturing and Computer Aided Engineering (CAD/CAM/CAE)—in particular—have been significantly influenced by several technological trends. These include deployment of cloud-oriented services for engineering and product data lifecycle purposes as well as the widespread availability of mobile IT devices.

The forthcoming large-scale availability of low-cost ubiquitous gesture tracking technologies, allowing for more natural interaction paradigms, together with the evolution of the computing scenario, will most likely have significant implications to 3D shape modelling.

2 An Overview of Recent and Current Trends for Shape Modelling in the Domains of Engineering and Manufacturing

2.1 From CAx to Virtual Engineering

As highlighted by a recent report by Jon Peddie Research, the domain of Computer-Aided technologies, often simply referred to as CAx, is "more dynamic than ever" [1] and can be accounted as the main responsible for a major evolution—within several engineering domains-towards Virtual Engineering (VE).

This evolution, which is often associated to the industrial uptake of Virtual Reality (VR) and Virtual Prototyping (VP) applications, is having profound implications across various design phases from concept to simulation and manufacturing [2]. A notable example of this evolution has been the all-virtual design of the Boeing 777 airplane [3, 4]. Such a milestone or achievement has been made possible through extensive use of CAx in the context of an extremely complex and collaborative set of design and simulation processes that involved experts, customers and manufacturers.

Several relevant projects funded by the European Commission within the fifth, sixth and seventh framework programs have specifically addressed Virtual Engineering. Some of the most relevant initiative include, but are not limited to, AIT VE-POP "Advanced Information Technology Virtual Early Prototyping Open Platform", VERDI "Virtual Engineering for Robust Manufacturing with Design Integration" and VEGA "Virtual Reality in Product Design and Robotics".

Despite the undoubtedly significant number of advantages, however, the use of VE and VR has also introduced a few critical drawbacks. One of the most notable examples is the false sense of security, conveyed by such an apparently deterministic approach. This clearly highlights the importance of developing modelling techniques that can intrinsically support variable levels of uncertainty as well as methodologies to formalize the degree of uncertainty across multiple modelling or processing/simulation steps.

The current limited awareness on the different degrees of uncertainty that characterized different components of a complex design solution may mislead to overconfident decisions made on top of only apparently precise datasets. Furthermore, as a result of the pervasive adoption of VE, the "fuzzy front end of the design process may be cut short—to the company's long-term disadvantage" [19].

Furthermore, the intrinsically high collaborative and interactive nature of these environments may lead to the dangerous—if uncontrolled-growth of number of changes with the consequent poor control over them. The potential impact of this factor has been dramatically demonstrated during the design of the A380 model airplane by Airbus [20], whose release to the market was significantly delayed due to the continuous changes of the design.

2.2 From Virtual Factories to Augmented Business Visualization

If we further expand the scope of our analysis, from the early stages of the design process to the whole manufacturing process, we can observe that the increasing complexity of manufactured products, e.g. in the automotive or aeronautics domain, requires high levels of flexibility, maintainability, and customization which is being increasingly addressed by Virtual Manufacturing (VM) systems.

The goal of VM is to pursue the integration of computer modelling for product development purposes within a manufacturing system. VM allows simulating all manufacturing operations to be performed on actual production lines and, as a discipline, it was borne out of the integration of many domains including CAD/CAM, Virtual Prototyping (VP), Virtual Reality (VR), production planning and control, Product Lifecycle Management (PLM), Product Data Management (PDM), Manufacturing Process Management (MPM), Computer-Integrated Manufacturing (CIM), Component Information System (CIS), Knowledge-Based Engineering (KBE), Manufacturing Process Planning (MPP), Electronic Design Automation (EDA).

Particularly relevant have been the implications emerging from the use of wide networked collaborative virtual manufacturing system [22] as in these cases the definition of product may require the cooperation between a large number of different experts using different tools. Referring to the aforementioned example of the design of the Boing 777 airplane, this required the definition of 130,000 parts, through the involvement of 6,800 internal and 10,000 external experts.

Furthermore, in the last few years, increasing use of Web Services and Grid computing has expanded the use of CSCD—Computer Supported Collaborative Design technologies. The importance of this trend in terms of shape modelling is significant as proved by the development of several commercial visualization systems for collaborative design, including Oracle AutoVue (formerly Cimmetry Systems), Actify SpinFire, Autodesk Streamline and SolidWorks DraftSight developed by Dassault Systèmes.

This trend is leading, as a result, to a new generation of applications that are designed for what is referred to as Augmented Business Visualization (ABV), which can be regarded as the application of Visual Analytics to the engineering and manufacturing domain.

2.3 The Importance of the Concept Design Phase

If we analyse the relevance of shape modelling research across the whole design and manufacturing lifecycle, we can observe that traditionally, significant effort has been paid to the development of visualization and visual computing technologies specifically designed for the early stages of the design process. In fact, particular attention has been paid to improving quality and efficiency of Computer Aided Styling (CAS) systems. The reason for this is to be found in the high market value of styling features in particularly important markets such as automotive or industrial design where products tend to be often remembered more for their aesthetical properties and "emotional features" than for their physical or mechanical properties. CAS software have been typically engineered to interactively support the designer along an evolutionary process that starts from a conceptual model or an initial—often vague—idea, which is iteratively refined until it is consolidated into a design "concept". Therefore, the main goal of CAS has been to free designers from very constrained graphical languages accounting, at the same time, for increasingly performing materials and manufacturing processes, which have allowed manufacturing of increasingly complex 3D shapes.

Virtual Prototyping (VP) tools, to a certain extent, have represented the natural evolution of CAS, in that they allow interactive simulation and optimization of products or concepts, assessment of design features through design reviews [5] or assessment of the impact on manufacturing processes. VP tools are now well established within advanced engineering fields, for instance within the automotive and aeronautical industries, and can allow considerable savings in the range of up to tens of millions of euros for each product.

In the past, several research groups have developed VP tools that leveraged on Virtual and Augmented Reality (VR/AR). The results were a number of three-dimensional environments that combined sketching and modelling within a natural seamless action. These environments allowed interactively creating and modifying 3D shapes at the early stages of the design process. First examples of these developments include the work of Butterworth et al. [6] and Steed et al. [7] while later works include those of Fiorentino et al. [8] or De Amicis et al. [9], to name but a few. More recent developments on interactive drawing of 3D models include, among others, studies at Fraunhofer IGD [12] on dynamic simplicial meshes [13], on physically based interactive simulation techniques for deformable objects Falcidieno and Léon [14], the work by Alexa [15], the work by Liverani [16] and the work by De Amicis [10].

2.4 *The "Massification" of Virtual Engineering*

Recent industrial trends in the domain of shape modelling have also been characterized by the need for mass-customization. It should be noted that the concept of "mass-customization" within this context could be considered from different standpoints.

The first "massification" trend sees the use of IT technologies to improve the design and manufacturing processes by extending the range of possible product configurations. With regard to this trend relevant works include the development of flexible vehicle manufacturing solutions and architectures that have been carried out, for instance, by the Smart Customisation group at MIT Media Lab [23].

The second approach to "massification" sees the possibility to create new IT-based services (especially in the context of web-services, for instance for marketing or commercial applications) that by themselves represent a customized form of product, resulting from a complex software engineering process. Examples of this second approach are the 3D services developed by My Virtual Model Inc., a Canadian company that allows creating virtual avatars of real people that can use their virtual counterparts to test or assess products (from garments to objects) in a virtual manner.

The third, and perhaps most challenging, approach to "massification" regards the massification of the design process itself through the involvement of large community of stakeholders, typically customers, within the design process, through very complex participated design process. This has profound implications in terms of shape modelling in that it requires development of complex participative interaction mechanisms and tools for collaborative shape creation and editing.

The fourth and last approach, which is also referred to as "consumerisation", sees the use of non-professional (consumer) applications being used for industrial or engineering use. The most notable example in the domain of shape modelling is the widespread use of the 3D modelling and visualization SketchUp package by Google for professional uses. Such a very significant (from a market standpoint) trend has been essentially fuelled by the interactive nature of these applications that offer a very easy-to-use interface at low cost and dramatically proves the importance of simplicity of use, within certain applications, even beyond the absolute quality of the final representation.

2.5 *The "Cloudification" Trend*

If we analyse the 3D modelling scenario from a different standpoint, the need for high-performance environments for 3D shape modelling and simulation has also driven the research community to explore the use of cloud-oriented software paradigms. An interesting example of this trend is the COVISE service-based platform [21], which extends the coSimLib interface for high-performance visualization for CPU/GPU systems for fast post processing tools, to allow interactive access through the web to results of 3D simulations.

Most notably, in fact, the result of the simulation can be visually manipulated through a WebGL client that renders a scenegraph created by a service-based post-processing tool on top of the results of the simulation. Although this approach does not yet allow for full simulation-steering, based on interactive manipulation of 3D shapes, it undoubtedly represents a significant step forward towards a fully interactive visualization and simulation environment.

Other examples of this trend include PythonOCC, a Python-wrapper for Open-CASCADE, which delivers CAD-based functions as a service—including visualization and simulation—through the web [20]. The resulting CAD/CAE web-based library introduces a logical layer between end-user-oriented CAD software, which provides end-user scripting and access to software-developer-oriented CAD kernels.

3 Existing Barriers and Opportunities

Based on the aforementioned trends it is possible to identify a number of barriers that need to be overcome by research in the next few years. If we start our analysis from the early stages of the product design lifecycle it clearly emerges how, in the medium-term, the role of tool superficially designed to support the styling phase is set to increase in terms of importance.

This will be mainly due to the wide availability of consumer technologies supporting more natural forms of interactions (e.g. multi-touch interaction technologies and low-cost motion sensing devices). The availability of low-cost consumer devices, instead of expensive customized setup used by research community and industry in the past few years, will eventually determine a rapid uptake of new interaction paradigms.

This is extremely important when we analyse the early stages of the design process, which is characterized by the use of CAS tools. Current CAS tools are in fact mainly designed for expert use and often make use on non-intuitive interaction process, for instance based on interactive manipulation of shapes through control points. However, users at this stage, typically designers, have limited mathematical knowledge of the underlying mathematical representation used by the system. For this reason it is difficult for them to acquire complete control overall shape definition process. This is particularly critical at the early stages of the design process when the operators needs very unconstrained ways to generate and modify shapes, for instance based on sketching and free-form modelling.

Among the several existing technological barriers to less constrained design process, few are worth particular attention. One the main barriers that limits the transition from physical to a totally virtual environments, is the development of experience augmentation technologies that can convey the sense of physical matter through interactive feedback mechanisms closely associated to the shape creation or editing action.

To this extent, new approaches based on emerging hardware (multi-touch devices, miniaturized portable gesture tracking systems) are required to fill the gap between virtual and physical scenes. Paradoxically, up to now, existing sketch-based modelling systems can be still considered at their infancy since they do not fully exploit the potential of emerging 3D or multi-touch/multi-hand interaction devices.

However bridging this gap is essential to help operators move beyond the level of usability achieved today by Computer Aided Styling (CAS) systems. This can be achieved through the development of both new mathematical models and novel forms of virtual feedback augmentation that can leverage on natural manipulation skills.

Traditionally, the use of multi-dimensional interaction and haptic devices, including motion capturing systems, gesture recognition technologies and haptic feedback devices, has been expensive, unpractical for desktop or mobile scenarios or has yielded unconvincing results. However, this is set to change rapidly, since the availability of low-cost, accurate and miniaturized feedback technologies will promote widespread adoption of 3D interactions at the consumer level.

This it is very likely to produce, as an effect, a cultural change whereby natural interactions will become part of the standard computer experience. As a result, this will eventually influence the development of more natural forms of interactions suitable for professional applications including those designed for the industrial domains.

These will have to move beyond past approaches that have only tried to mimic traditional techniques such as clay modelling. In fact it is very likely that in the near future, due to the flexibility allowed by forthcoming gesture tracking technologies (e.g. concurrent tracking of multiple users), mockup-modelling techniques such as clay modelling and sketching will play an increasing smaller relevance.

The availability of more natural interfaces within the product development process could shorten the product development time and improve the overall "quality" of industrial processes and products. The concept of "quality" here is regarded in very general terms, ranging from the quality of manufacturing processes (in terms of efficiency, safety and ergonomics) as perceived by the industry and by those working therein, to the quality of product life cycle (in terms of maintenance) up to the quality of the product itself (in terms of aesthetics, usability, ergonomics and functioning) perceived by the final users.

For this to happen it will be necessary not only to address more efficient interaction mechanisms but also new cooperation mechanisms to allow on interactive shape modelling by several concurrent users. This will require development of forms of concurrent modelling within a three-dimensional realistic environment, for instance to support cooperation and evaluation by operators from different sites, both inside and outside the boundaries of a company. This would allow more collaborative enterprises environment leveraging on knowledge coming from suppliers, whose operator would be thus able to better validate a product before it is physically realized.

The main challenge will be to deliver technologies characterized by extremely high usability based on very natural forms of interaction and dialogue with the system. This will also have implications in terms of support to mass customization in that it will help non-experts (e.g. final users) to model and/or modify 3D shapes of virtual products in a simple way. This would allow for instance final user to be able to participatively involved, through support of methods such Analytic Hierarchy Process—AHP or Multi-criteria decision analysis—MCDA, in the assessment and proactive designs of products.

Moving to a different issue, significant limitations still exist today when considering interactive and iterative tools addressing the real-virtual-real object transformation cycle. It is a matter of fact that the use of digitization techniques today is still not fully integrated with the virtual manufacturing pipeline. Alternatively, its extensive use may be unpractical in that it may either require significant manual work (for instance when scanning the interior of a vehicle) or time (for instance when acquiring the exterior of a vehicle).

Furthermore the use of 3D digitization systems requires significant post-processing and manual work to create a model from the initial point cloud or mesh based on transformations and re-modelling to ensure that the final shape can be further modified in an interactive manner.

A classical example of this can be appreciated when observing a point cloud resulting from a scanning process. The point cloud cannot be modified in an intelligent or semantically meaningful manner. Therefore its practical use within an interactive design process requires a time-consuming manual process to transform the dataset from a discrete to a continuous representation. This example clearly highlights the need for the development of new high-level shape models that can integrate, from a user perspective, both continuous and discontinuous representation. It is therefore clear that a significant barrier to be removed is the development of unified representation over the different incompatible or only partially compatible computer representations used today, which often hinder several operational activities.

Furthermore, future tools for shape re-engineering will require, for instance, new algorithms for shape semantic understanding and interpretation based on machine learning technologies, computer vision, swarm of communication systems and 3D descriptors, to allow for easy modification and account for downstream processes. The availability of these processes would be very beneficial to help users define hypothesis about the initial data structure, thus accelerating the transformation towards more usable representation.

In more general terms, it can be stated that there is a clear need for semantically rich 3D model descriptions that can account for intrinsic knowledge representations to facilitate both generation and transformation of 3D models in order to ensure that the representation used within a given context is best fit-for-the-purpose. To do so the research community will have to address the development of novel representation schemas such as unified models for NURBS, sub division schemes, locally refined splines, integrating semantics as an intrinsic feature of the geometric descriptions.

The development of more knowledge-rich models, in turn, would foster the creation of more intelligent tools to automatically facilitate manipulation or transformation tasks or to provide intelligent control mechanisms that do not require specific expert knowledge. This, in turn, would allow development of more intelligent systems capable to encourage use by non-experts, thus contributing to the overall usability of design systems as discussed earlier within this section.

This issue becomes particularly relevant when observing that CAD or CAE models are only limitedly adopted outside the design and manufacturing stages, for instance for advertisement, communication activities or to create interactive operating manuals. This is essentially caused by the high complexity of CAD/CAE models as well as by the need to create graphical effects capable to express "emotional" features that are not defined within the original CAD dataset. Having specific models fit-for-the-purpose (in this case for aesthetic rendering) automatically created from an initial model would be of significant industrial value. Additionally, the exchange of properties across different CAD environments is far from being sorted fully and often relies either on use of proprietary exchange formats, such as IGES, Parasolid XT and JT Open, or on use of intermediary neutral formats, most notably STEP, with a number of limitations.

The evolution towards more "intelligent" shapes should leverage on new generation of Computer Aided Innovation (CAI) tools that will have to help innovators, designers, engineers better express their creative potential across the overall product development lifecycle, and not only limitedly to the early stages of the design process.

For this reason it will be necessary to introduce innovative paradigms that allow creative processer over a unified representation, at higher abstraction level than CAD, multi-physics CAE and Product Lifecycle Management (PLM) systems. As a result, feature-based approach should be closely integrated with functional and multi-physics analysis typical of CAE environments. This would help extend the scope of interactive shape creation and modification from a merely geometrical standpoint to a function-centred paradigm.

This evolution will pave the way to the creation of automatic mechanisms to handle intelligent shapes that will support consistent variations across multi-physics environments, based on semantics rather than only on 3D geometrical description. It is important to underline that this evolution will also have implications at the manufacturing level in that it will allow development of more intelligent machinery capable to operate based on interpretation and understanding of the true semantic meaning of 3D shapes.

This vision eventually highlights one of the highest barriers that need to be removed, that is the existence of very different data models between by the various vertical "families" of software used across the various design and manufacturing stages. Furthermore, it should be noted that similar differences also exist horizontally across "families" of software tools used within different industries, namely:

- CAx systems, essentially responding to requirements of manufacturing and production engineering.
- Building Information Models (BIM), which have been developed to comply with requirements emerging from architecture and engineering domain.

- Geographical Information System (GIS), which have been driven by the requirements of urban planners, service engineering, natural scientists and security experts.

Traditionally, each of the aforementioned software family has developed—in a fairly independent manner—different internal data representation models. This today significantly limits development of cross-domain tools. Furthermore, the lack of a unified low-level representation has produced data silos that are cause of limited synergies and inefficiencies.

Removing such a barrier would bring significant industrial and market benefits to a range of engineering domains, allowing for cross-domain integration without being forced to rely on technology-neutral intermediate data translations.

4 A Forward-Looking Analysis and Conclusions

Starting from the aforementioned analysis, it is possible to identify a future scenario that may unfold in the next few years by analysing existing barriers and technological trends, and by projecting the development of current scientific achievements of relevance for domain of 3D shape creation and modelling in the context of industrial applications.

If we follow this approach, we can predict that the medium-term future research activities may contribute to the convergence of traditional Modelling and Simulation (M&S) systems, as we know them today, into a unique real-time interactive user-centric environment that will be used for integrated modelling, simulation (including "co-simulation") and visualization.

For this reason, research will have to address how to let users appreciate aesthetic aspects of a 3D shape while interacting with it according to specific physical properties, and receiving real-time feedback on the effects of the modifications in terms of functionalities, ergonomics or structural properties.

Such shape modelling/simulation algorithms will have to maximize the benefits in terms of computing power emerging from forthcoming computing hardware and architectures. The latter will be increasingly characterized by very heterogeneous computing ecosystems leveraging on CPUs and GPUs and by very high number of cores, up to thousands. Future shape modelling/simulation algorithms will also have to maximize benefits of new computing paradigms, based for instance on quantum or DNA computing, or new computing materials, for instance grapheme or silicene, which may allow manufacturing of processors running at terahertz speed and beyond and which may soon bring to unprecedented high-performing computing hardware.

This will require research to address a number of scientific challenges among which the most relevant ones are:

- Development of interactive, user-friendly and semantically meaningful modelling of shapes.

- Development of interoperable representation scheme bridging existing gaps among different continuous and discrete data structures.
- Development of real-time interactive simulation of the behaviour of 3D shapes (e.g. structural, etc.) leveraging on use of emerging heterogeneous computing architectures.
- Development of interactive co-simulation methods closely linked to the digital modelling stage to ensure integration of geometrical features with functional or physical constraints.

To conclude, as noted by several authors, "product design and manufacturing improvements drove research from the 1960s through the 1980s" [22]. However, thanks to the technical development, industries have increasingly used the available, and progressively more effective, technologies, lowering interest for researching new solutions [22]. Although progress brought by academic research have been significant, however, the fast changing scenario brought by technological evolution and the key impact on production and industrial process that this may bring, should suggest industry to refocus once more on research in the domain of shape modelling.

References

1. Jon Peddie Research: Jon Peddie Research releases the Worldwide CAD Market Report 2012. 2012. http://jonpeddie.com/press-releases/details/worldwide-cad-market-report-2012/ (2012). Accessed 18 February 2013
2. Vornholt, S., Geist, I., and Li Y: Categorisation of data management solutions for heterogeneous data in collaborative virtual engineering. In: Proceedings of the First International Workshop on Digital Engineering (IWDE '10), ACM Press, New York, NY, USA. (2010)
3. Sabbagh, K.: Twenty-First Century Jet: The Making and Marketing of the Boeing 777. Scribner, New York (1996)
4. Marion, T., Fixson, S., and Meyer, M.H.: The Problem with Digital Design. MITSloan Manag. ReV. **53**(4), 63–68 (2012)
5. Dunston, P.S., Arns, L.L., and McGlothin, J.D.: An Immersive Virtual Reality Mock-Up for Design Review of Hospital Patient Rooms. In: Proceedings of the 7th International Conference on Construction Applications of Virtual Reality. University Park, Pennsylvania, October 22–23 (2007)
6. Butterworth, J., Davidson, A., Hench, S., and Olano, T.M.: 3DM: A Three Dimensional Modeler Using a Head-Mounted Display. In. Communications of the ACM, pp. 55–62, June (1992)
7. Steed, A., Slater, M.: 3D interactions with the desktop bat. Computer Graphics Forum, Blackwell. **14**(2), 97–104 (1995)
8. Fiorentino, M., De Amicis, R., Stork, A., and Monno, G.: Surface Design in Virtual Reality as Industrial Application". In: Marjanovic (Ed.) Proceedings of the Design 2002–7th International Design Conference. Dubrovnik, Croatia, pp. 4 477–482. The Design Society, Glasgow (2002).
9. De Amicis, R., Bruno, F., Stork, A., and Lucchi, M.L.: The Eraser Pen: a New Interaction Paradigm for Curve Sketching in 3D. In: Marjanovic (Ed.) Proceedings of the Design 2002–7th International Design Conference. Dubrovnik, Croatia, pp. 465–470. The Design Society, Glasgow (2002)
10. De Amicis, R., Fiorentino, M., Santos, P., and Stork, A.: SketchAR - sketching in mixed realities. In: Gausemeier J. (ed.) Augmented und Virtual Reality in der Produktentstehung, pp. 145–156. HNI-Verlagsschriftenreihe (2002)

11. Conti, G., and Ucelli, G.: A Java3D^{TM} Tool for Real-Time Collaboration in a Virtual Reality CAAD Environment. In: Proceedings of Design and Decision Support Conference, Ellecom, The Netherlands (2002)
12. Graf, H., Peña Serna, S., and Stork, A.: Towards conceptual simulations within highly interactive environments. In: Proceedings of International Conference on Modeling, Simulation & Visualization Methods (MSV 2007) Las Vegas, pp. 108–114. CSREA Press (2007)
13. Peña Serna, S., and Stork, A.: Dynamic simplicial meshes. In: Proceedings of SIGGRAPH '09: Posters (SIGGRAPH '09). ACM, New York, NY, USA. Article 64, 1 (2009)
14. Cheutet, V., Catalano, C.E., Pernot, J.-P., Falcidieno, B., Giannini, F., Léon, J.-C.: 3D sketching for aesthetic design using fully free-form deformation features. Computers & Graphics. **29**(6), 916–930 (2005)
15. Nealen, A., Sorkine, O., Alexa, M., Cohen-Or, D.: A sketch-based interface for detail-preserving mesh editing. In: Markus Gross, (Ed.) Proceedings of ACM SIGGRAPH 2005 Papers (SIGGRAPH '05), pp. 1142–1147. ACM, New York NY, USA (2005)
16. Liverani, A., Ceruti, A., and Caligiana, G.: Tablet-based 3D sketching and curve reverse modelling. International Journal of Computer Aided Engineering and Technology, special Issue on: Digital Human Modelling and Tracking Human Actions (2012)
17. Liverani, A., Ceruti, A., and Caligiana, G.: A Spatial Tracked Touch Screen for Computer Aided Sketching. In: Proceedings of TMCE 2010 Symposium - Eighth International Symposium on Tools and Methods of Competitive Engineering. Delft University (Ed.), Ancona, Italy, 1(2), pp. 621–630 (2010).
18. Wong K.: What Grounded the Airbus A380? Cadalyst MCAD Tech News 194. Dec. 7, 2006. http://www.cadalyst.com/manufacturing/news/mcad-tech-news-194-11900. Accessed 18 February 2013
19. Niebling, F., Kopecki, A., and Becker, M.: Collaborative steering and post-processing of simulations on HPC resources: everyone, anytime, anywhere. In: Proceedings of the 15th International Conference on Web 3D Technology (Web3D '10), pp. 101–108. ACM, New York, NY, USA (2010)
20. Paviot, T., and Feringa, J.: Implementation of a SaaS Based Simulation Platform Using Open Standards and Open Source Software. In: Proceedings of the 12th NASA-ESA Workshop on Product Data, Exchange (PDE2010) (2010)
21. Nidamarthi, S., Allen, R.H., Sriram, R.D.: Observations from supplementing the traditional design process via Internet-based collaboration tools. International Journal of Comput Integr Mfg **14**, 95–107 (2001)
22. Kasik, D.J.: The Third Wave in Computer Graphics and Interactive Techniques. Computer Graphics and Applications, IEEE, 31(4), pp. 89–93, July-Aug (2011). doi:10.1109/MCG.2011.64
23. MIT Smart Customization Group. http://scg.mit.edu/ Web. 2 December 2013

An Hilbert-Driven Strategy for Signature-Based Gröbner Basis Algorithms

Bruno Simões

Abstract Gröbner bases are special sets of polynomials with nice properties that can be used to solve many problems in mathematics, natural and technical sciences. Examples of applications in geometric and solid modeling include intersecting surfaces and curves, finding closest point on a curve or on a surface to the given point, finding equations for equidistant curves and surfaces to curves and surfaces defined in terms of polynomial equations, etc. One of the major limitations of the algorithms to compute Gröbner bases is their computation time. We present a new criterion to improve the performance of Gröbner bases algorithms, driven by polynomial signatures. This criterion relies on properties of the Hilbert series and on the information gathered about the syzygy module of the ideal, to discard useless critical pairs.

Keywords Gröbner basis · Hilbert series · Signature-based Strategy

Mathematics Subject Classification (2010) 08A40 · 11Cxx · 12-XX · 13P10 · 11D72

1 Introduction

Geometric and solid modeling have received great attention from the academic and industrial communities throughout the past four decades, and still they can intrigue us with their issues concerning the representation and manipulation of objects. Many of the geometric formulations in these two fields, such as curves and surfaces, can be represented in terms of polynomial equations. Such algebraic representations are also useful to describe their problems and to derive a solution e.g. boundary computations.

B. Simões (✉)
GraphiTech Foundation, Via alla Cascata, 56/c, 38123 Povo, Italy
e-mail: bruno.simoes@graphitech.it

R. De Amicis and G. Conti (eds.), *Future Vision and Trends on Shapes, Geometry and Algebra*, Springer Proceedings in Mathematics & Statistics 84,
DOI: 10.1007/978-1-4471-6461-6_2,

The task of solving polynomial systems, especially of geometric constraints, is therefore a fundamental requirement in many applications e.g. mechanical assembly planning, tolerance analysis, constraint-based sketching and design, kinematic analysis of robots, and other mechanisms. An important class of problems involves finding the orientations, positions and dimensions of a set of geometric entities, which have to satisfy a set of geometric constraints. Earlier approaches to solve large constraint systems of this type have relied on semi-automatic procedures but automated approaches based on problem modeling have emerged recently in the literature, thereby opening doors to a larger set of unsolved problems and stressing the need for extremely efficient algorithms. The necessity of solving algebraic equations is also present in surface intersection algorithms, collision detection, ray tracing of curves and surfaces, manipulation of offsets of curves and surfaces [13] and geometric theorem proving [12, 15].

Techniques for solving non-linear polynomial systems can be categorized into geometric, symbolic and numeric methods. Symbolic methods based on algorithms for computing Gröbner bases or Resultants can be used for eliminating indeterminates and thereby reducing problems to a search of polynomials roots in the univariate case. Also, methods based on the computation of Gröbner bases have, for instance, other interesting properties such as a practical approach to check whenever a polynomial system has solutions or not. These techniques are an extraordinary example of how an idea used to solve one problem can become the key for solving a large variety of other problems, including in different areas. However, existing algorithms are known to be efficient only for polynomial systems of low degree. One of the major problems arises from the fact that finding the roots of univariate polynomials can be ill-conditioned for polynomials of degree greater than 14, as shown by Wilkinson [19]. As a result, the implementation of these algebraic methods with finite arithmetic precision is considered to be challenging as it constrains the performance of the resulting algorithms. Nevertheless, the prevailing viewpoint is that using algebraic methods in geometric and solid modeling usually leads to a better theoretical understanding of the problems. Therefore, they can be used to uncover new opportunities in the upcoming paradigms of smart geometry, but at the moment, their practical impact is still not clear.

Numeric methods can be sub-categorized into homotopy and iterative methods. The performance of the homotopy methods can be quite reasonable but it usually requires a solid theoretical background of the subject. Homotopy methods based on continuation techniques rely mostly on path-following techniques in the complex space. In theory, each path being followed should converge to a geometrically isolated solution but in practice we have to deal with additional issues, for example, the different paths being followed might not be geometrically isolated.

Iterative methods, e.g. Newton's method, are procedures to compute a sequence of improving approximate solutions to the problem. They are most suitable for local analysis and they work well when provided with a good initial guess of the solution. Though, supplying such a good guess in applications such as geometric constraint systems or intersections is not a trivial task. Recently, iterative methods based on interval arithmetic have received a considerable amount of attention in computer graphics and geometric modeling. The resulting algorithms are robust, though their convergence can sometimes be relatively slow.

This paper is devoted to efficient computations of Gröbner basis, with a particular emphasis on signature-based attempts. Our main result is based upon the idea that we can extend signature-based attempts by using the properties of the Hilbert series, and improve its computational efficiency.

The structure of this paper is the following. In Sect. 2, we recall the basic concepts and terminology. In Sect. 3, we show that improvements to the state-of-the-art are still possible by taking advantage of the data collected by signature-based techniques. In Sect. 4, we benchmark some of the fastest algorithms to compute a Gröbner basis against a standard database of examples, and we assess the performance of our algorithm. We shall see that it provides good results. We conclude this paper with a short list of remarks and future directions regarding the current results.

2 The Theory of Gröbner Bases

As often happens, the invention of the theory of Gröbner bases can be associated to several mathematicians. In our view, the major step was taken by Bruno Buchberger during his PhD thesis [3]. First, he formulated the concept of Gröbner bases, extending a suggestion of his advisor Wolfgang Gröbner. Such influence is also acknowledged with his terminology. Lastly, he found an algorithm to compute them, as well as a proof for the fundamental theorem on which the correctness and termination of the algorithm hinges.

This section attempts to present in a self-contained manner the basic facts about the theory of Gröbner basis. We start with a short introduction to polynomial rings and their ideals. Then, we bring the reader into the realm of orderings. Term orderings are an important tool because they enable us to write polynomials in a well-defined way that can then be implemented on a computer. After ordering terms in polynomials or tuples of polynomials completely, their leading terms can be singled out and then used to build leading term ideals and modules, which are conceptually much simpler to handle. Then we review the theory of Hilbert series, which is one of the concepts required to fully understand the proposed criterion. Lastly, we introduce the theory of Gröbner bases and the signature-based strategy.

For a more detailed introduction on commutative algebra we refer the reader to [1, 6]. Good books covering those topics with a more detailed introduction to the theory of Gröbner bases and a stronger emphasis on computational aspects are, for example, [14, 15]. Most of the proofs given in this section are either easy or can be found in any introductory book about commutative or computational algebra (for example in the ones mentioned above). We focus ourselves on the theory of Gröbner bases, thus we provide only proofs if they are short, beautiful, and give some deeper insight on the topics covered, otherwise we provide only references.

2.1 Theoretical Foundations

We assume that the reader has some basic knowledge of basic algebra, but we do not think it is harmful if we recall some fundamental definitions. Also, readers familiar with this topic may want to skim this section for notation and terminology.

Let P $= R[x_1, \ldots, x_n]$ be a polynomial ring over a ring R that consists of the multivariate polynomials of the form:

$$\sum a_{\alpha_1,\ldots,\alpha_n} x_1^{\alpha_1} \ldots x_n^{\alpha_n}$$

where $a_{\alpha_1,\ldots,\alpha_n} \in R$ and α_i are non-negative integers. Then we have the following definition.

Definition 1 A *term* (or a *power product*) is a product of n indeterminates, each raised to the power of a non-negative integer. It can be written concisely as

$$x^\alpha = x_1^{\alpha_1} \ldots x_n^{\alpha_n} \tag{1}$$

where α is the coordinate vector $(\alpha_1, \alpha_2, \ldots, \alpha_n) \in \mathbb{N}^n$.

Definition 2 The *total degree* of a term x^α is simply defined as $\deg(x^\alpha) = |\alpha| = \alpha_1 + \alpha_2 + \cdots + \alpha_n$. The degree of x^α in any indeterminate x_i is $\deg_{x_i}(x^\alpha) = \alpha_i$.

We denote the set of all terms involving the indeterminates $x_1, \ldots, x_n$ by the expression $\mathbb{T}^n$. Also, we say that a term $p = x_1^{\alpha_1} \ldots x_n^{\alpha_n} \in \mathbb{T}^n$ is *divisible* by $q = x_1^{\beta_1} \cdots x_n^{\beta_n} \in \mathbb{T}^n$ if $\beta_i \leq \alpha_i,\ \forall\, 1 \leq i \leq n$. The definition of the least common multiples and the greatest common divisors of two terms is the following.

Definition 3 Let P $= K[x_1, \ldots, x_n]$ be a polynomial ring over a field K. Let $t_1 = \prod_{i=1}^n x_i^{\alpha_i} \in \mathbb{T}^n$ and $t_2 = \prod_{i=1}^n x_i^{\beta_i} \in \mathbb{T}^n$ be two terms. Then we define

$$\mathsf{lcm}(t_1, t_2) = \prod_{i=1}^n x_i^{\max\{\alpha_i, \beta_i\}} \quad \text{and} \quad \gcd(t_1, t_2) = \prod_{i=1}^n x_i^{\min\{\alpha_i, \beta_i\}}$$

as the *least common multiple* of t_1 and t_2, and the *greatest common divisor* of t_1 and t_2, respectively.

We say that t_1 and t_2 are *co-prime* or relatively prime if $\gcd(t_1, t_2) = 1$.

Definition 4 The *support* of a polynomial f is the set $\mathrm{Supp}(f) = \{x^\alpha \in \mathbb{T}^n \mid \mathrm{coef}(c_\alpha x^\alpha) \neq 0\}$. A polynomial f is said to be *homogeneous of degree d* if all terms have the same total degree d.

Another important definition is the one called *monomial*.

Definition 5 A *monomial* is an element of P of the form $m = ct$, where $c \in R$ is its *coefficient* and $t \in \mathbb{T}^n$ is its *term*. The *total degree* of the monomial m is simply the total degree of t.

Next we show that ideals generated by terms are finitely generated. The importance of such result is self-evident in the next sections. However, it should be already clear that statements about "finiteness" are extremely valuable from a computational point of view.

Definition 6 A *monoid* $(\Gamma, +)$ is a set Γ together with an associative operation + on it, such that there exists an identity. A monoid whose operation is commutative is called a *commutative monoid*.

In this paper, the operations are always considered to be commutative, and we shall use the term "monoid" to denote a commutative monoid.

Definition 7 Let $(\Gamma, \cdot)$ be a monoid. A non-empty subset $\Delta \subseteq \Gamma$ is called a *monomial ideal* in Γ if the following property holds: $\Delta \cdot \Gamma \subseteq \Delta$.

Proposition 1 *The following conditions are equivalent for a monoid* $(\Gamma, \cdot)$.

1. *Every monomial ideal in Γ is finitely generated.*
2. *Every ascending chain $\Delta_1 \subseteq \Delta_2 \subseteq \ldots$ of monomial ideals in Γ is eventually stationary.*
3. *Every non-empty set of monomial ideals in Γ has a maximal element with respect to inclusion.*

The monoid Γ is said to be Noetherian if these conditions are satisfied. The same claim is valid for the monoid $(\mathbb{N}^n, +)$.

Proof See, for example, Sect. 1.3 in [14].

This proposition provides an important finiteness condition for ideals in polynomial rings.

Corollary 1 (Dickson's Lemma) *Every set of terms $X \subseteq \mathbb{T}^n$ contains a finite subset $Y \subseteq X$ such that each $p \in X$ is a multiple of some term in Y.*

Proof Let log:$\mathbb{T}^n \to \mathbb{N}^n$ be the map given by $x_1^{\alpha} \ldots x_n^{\alpha} \to (\alpha_1, \ldots, \alpha_n)$. Such map is clearly an isomorphism of monoids. Hence, the monomial ideal $(\log(t_1), \log(t_2), \ldots) \subseteq \mathbb{N}^n$ is finitely generated by the previous proposition, and there exists a number $i > 0$ such that this monomial ideal is generated by $\log(t_1), \ldots, \log(t_i)$ where $t_i \in \mathbb{T}^n$. Therefore, we have that the set of terms $\{t_1, \ldots, t_i\}$ generates the monomial ideal $(\log(t_1), \log(t_2), \ldots) \subseteq \mathbb{T}^n$.

The generalization of the Dickson's Lemma is given by the following theorem.

Theorem 1 (Structure Theorem for Monomial Modules) *Let $M \subseteq \mathrm{P}^r$ be a monomial module.*

1. *The module M is finitely generated. That is, $M = \langle t_1 e_{\alpha_1}, \ldots, t_s e_{\alpha_s} \rangle$, where s is finite, $t_1, \ldots, t_s \in \mathbb{T}^n$ and $\alpha_i \in \{1, \ldots, r\} \subset \mathbb{N}$*
2. *There are monomial ideals $I_1, \ldots, I_r \subseteq \mathrm{P}$ such that M is of the form $M \cong \oplus_{i=1}^r I_i e_i$.*

Proof Assume $B \subseteq \mathbb{T}^n \langle e_1, \ldots, e_r \rangle$ to be a system of generators of the monomial module M. Let $B_i = \{t \mid te_i \in B\} \subseteq \mathbb{T}^n$ be a set of terms with $1 \leq i \leq r$. By Dickson's Lemma, the monomial ideal $I_i = (B_i)$ has a finite systems of generators $S_i \subseteq B_i$. Clearly, the P-module M is then generated by $S_1 e_1 \cup \ldots \cup S_r e_r \subseteq \mathbb{T}^n \langle e_1, \ldots, e_r \rangle$. This proves the finiteness and also the claim that $M = \sum_{i=1}^r I_i e_i$. The fact that this sum is direct follows from $M \subseteq \oplus_{i=1}^r \mathrm{P} e_i$.

There is an analogy between Theorem (1.1) and Proposition (1.1) that implies that the analogue of Proposition (1.2) holds for monomial modules as well.

Corollary 2 *Every ascending chain of monomial submodules of* P^r *is eventually stationary.*

Proof Let $M_1 \subset M_2 \subset \ldots$ be a strictly ascending chain of monomial submodules of P^r, and assume that such chain exists. Then we have that $\langle t_1, \ldots, t_i \rangle \subseteq M_i$ for all $i \geq 1$ because these are all modules generated by terms. Assume that we can find for every $i \geq 1$ a $i + 1$ such that $t_{i+1} \notin \langle t_1, \ldots, t_i \rangle$. Then the monomial submodule $\langle t_1, t_2, \ldots \rangle$ of P^r is not finitely generated. This represents a contradiction to Theorem 1.

Lastly, we shall see the uniqueness for systems of generators of monomial modules.

Proposition 2 *Assume $M \subseteq P^r$ to be a monomial submodule. Then there exists a unique minimal element in the set of all systems of generators of M with respect to inclusion. We denote such set as the minimal monomial system of generators of M.*

Proof See, for example, Sect. 1.3 in [14].

One of the questions that remains unanswered is: in how many different ways can we write a polynomial $p \in \mathrm{P}$? Clearly, the terms in the support of a polynomial can be ordered by decreasing or increasing degree. However, that is not enough since different terms can have the same degree. Our next task is to equip them with a certain additional property called a term ordering, which allow us to define them uniquely. This fundamental property is the key for the finiteness of most algorithms we shall encounter later.

In the univariate case the operation is straightforward because we can define $x^a \succ x^b$ as being true if and only if $a > b$. But for the multivariate case we need to find a new relation that is capable of providing such uniqueness and that is consistent with polynomial multiplication.

Definition 8 A *total ordering* is a set S and a relation on that set that satisfies the conditions for a partial ordering plus an additional condition known as the comparability condition. A relation $\leq$ is said to be a total ordering on a set S if the following properties hold.

- Reflexivity: $a \leq a$ for all $a \in S$.
- Antisymmetry: $a \leq b$ and $b \leq a$ implies $a = b$.
- Transitivity: $a \leq b$ and $b \leq c$ implies $a \leq c$.
- Comparability (trichotomy law): For any $a, b \in S$, either $a < b, a > b$ or $b = a$.

The first three axioms define only a *partial ordering*, whereas the addition of the trichotomy law defines a total ordering. Every totally ordered set that is finite is well ordered.

Definition 9 A *well-ordering* on a given set is a total ordering such that any non-empty subset has a smallest element w.r.t $\succ$.

Definition 10 An *ordering* $\succ$ is an *admissible term ordering* if it satisfies three conditions:

1. It is a total ordering on the set of all terms $\mathbb{T}^n$.
2. It is multiplicative; i.e., $x^a \succ x^b$ implies $x^{a+c} \succ x^{b+c}$ for all $a, b, c \in \mathbb{N}^n$.
3. It is a well-ordering, namely, the constant term is the smallest.

Term orderings have other useful properties. Once we fix an ordering, every polynomial f has a unique leading term. Moreover, that unique leading term is the largest term in the polynomial (w.r.t. that term ordering) with a nonzero coefficient in the expansion of f.

Definition 11 Let $f = c_\alpha x^\alpha + \sum c_k x^{\beta_k} \in \mathrm{P}$ such that $x^\alpha \succ x^{\beta_k}$ and $c_\alpha \neq 0$. Then:

- $\mathsf{LC}_\succ(f) = c_\alpha$ is the *leading coefficient* of f.
- $\mathsf{LT}_\succ(f) = x^\alpha$ is the *leading term* of f.
- $\mathsf{LM}_\succ(f) = \mathsf{LC}_\succ(f) * \mathsf{LT}_\succ(f) = c_\alpha x^\alpha$ is the *leading monomial* of f.

Up to now, we saw how to define polynomials uniquely in P. Next we extend the definition of term orderings on polynomial rings to the one on monomial modules. There are two reasons that can justify our interest in such orderings. First we are interested in the computation of Gröbner bases of modules. Lastly, signature-based algorithms work with elements of such modules. Hence, we start with a review of the Definition 1.

Definition 12 Assume M to be a free P-module of rank r with canonical basis elements e_i. A *module term* in M is an element of the form $t = x^\alpha e_i$ where x^α is a term in P. The set of all module terms is denoted $\mathbb{T}^n \langle e_1, \ldots, e_r \rangle$.

The set of all module terms $\mathbb{T}^n \langle e_1, \ldots, e_r \rangle$ is obviously a disjoint union of r copies of $\mathbb{T}^n$, where e_1, ..., e_r merely indicate the copy we are considering.

Definition 13 Let $x^\alpha e_i$ and $x^\beta e_j$ be two module terms in $\mathbb{T}^n\langle e_1, \ldots, e_r\rangle$. We say that $x^\alpha e_i$ divides $x^\beta e_j$ if and only if $i = j$ and $x^\alpha \mid x^\beta$, for all $x^\alpha, x^\beta \in \mathbb{T}^n$. We use $x^\alpha e_i \mid x^\beta e_j$ as a shorthand notation for this operation.

Before we move into the realms of module orderings, we shall revise a few other properties of module terms.

Definition 14 Let M to be a free P-module of rank r with canonical basis elements e_i, and R be a ring. The expression term shall denote a module term if no confusion arises. Hence, it can be said

1. A *monomial* $m = ct$ in M is a term $t \in M$ times a coefficient $c \in R$.
2. The *index of a term* $t = x^\alpha e_i$ is denoted $index(t) = i$.
3. Computing the *degree of a term* $m = x^\alpha e_i$ can be reduced to the one of the term $x^\alpha \in \text{P}$, as defined in Definition 4: $\deg(m) := \deg(x^\alpha) = \sum_{i=1}^{n} \alpha_i$.

We can write a element $f \in M$ as a sum of monomials

$$f = \sum_{i=1}^{r} \left(\sum_{\alpha \in \mathbb{N}^n}^{\text{finite}} c_\alpha x^\alpha \right) e_i$$

such that $c_\alpha \in R$ and $x^\alpha \in \mathbb{T}^n$. However, this representation is again unique only up to the order of the monomials. Thus, a term ordering on M is also required. Naturally, this is just a generalization of a term ordering on P that takes into account the canonical basis elements e_i,

Definition 15 Let $\succ_\sigma$ be a term ordering on P and $M = \oplus_{i=1}^{r} = \text{P}e_i$ a free P-module of rank r with canonical basis elements e_i. A *module term ordering* $\succ$ is a compatible total ordering on the set of all terms of M such that $x^\alpha e_j \succ x^\beta e_j \Rightarrow x^\gamma x^\alpha e_i \succ x^\gamma x^\beta e_j$ and $x^\alpha \succ_\sigma x^\beta \Rightarrow x^\alpha e_i \succ x^\beta e_i$, for all $\alpha, \beta, \gamma \in \mathbb{N}^n$ and $i, j \in \{1, \ldots, r\}$.

Clearly, a term ordering on the polynomial ring P can also be understood as a module term ordering on the module $\text{P} \cong \text{P}e_1$. Therefore, the last definition is just a generalization of the Definition 10.

Definition 16 Given a module ordering $\succ$ on M, every element $f \in M$ can be uniquely represented by $f = c_\alpha x^\alpha e_i + f'$, and for all nonzero terms $c_\beta x^\beta e_j$ of f' it holds that $x^\alpha e_i \succ x^\beta e_j$.

Once we fix a module ordering, every polynomial f has the following properties.

Definition 17 Let f be a polynomial in a module M.

- the leading term of f is $\mathsf{LT}_\succ(f) = x^\alpha e_i$,
- the leading coefficient of f is $\mathsf{LC}_\succ(f) = c_\alpha$,
- the leading monomial of f is $\mathsf{LM}_\succ(f) = c_\alpha x^\alpha e_i$,

Likewise in the polynomial case, we shall always equip a module M with a module term ordering $\succ$. Hence, the use of the notation M implies also the use of a module term ordering on M.

Next we review a few more properties concerning ideals and modules such as the relation they have with their leading term ideals and modules. These properties will come in handy once we need an answer to questions like: Let I be an ideal in a polynomial ring $\mathrm{P} = K[x_1, \ldots, x_n]$ over a field K. Obviously, the residue class ring P/I can be viewed as a K-vector space. So, is it possible to exhibit an explicit basis?

We shall see that with the help of leading terms, the Macaulay's Basis Theorem yields a noteworthy answer to this question. This theorem is based on the assumption that the module ordering is a term ordering and the base ring is a field. That underlines for the first time the theoretical importance of term orderings.

Definition 18 Let $M \subseteq \mathrm{P}^r$ be a P-submodule and $\succ$ fixed term ordering.

1. The module $\mathsf{LT}_\succ(M) = \langle \mathsf{LT}_\succ(m) \mid m \in M \backslash \{0\} \rangle$ is said to be the *leading term module* of M with respect to $\succ$.
2. If $M \subseteq \mathrm{P}$ then we say that $\mathsf{LT}_\succ(M) \subseteq \mathrm{P}$ is the *leading term ideal* generated by the leading terms of all the elements in M with respect to $\succ$. A term x^α is called *standard* if it does not belong to $\mathsf{LT}_\succ(M)$.

Assume that $m_1, \ldots, m_s \in \mathrm{P}^r$ are non-zero vectors and that $M = \langle m_1, \ldots, m_s \rangle \subseteq \mathrm{P}^r$ is the submodule generated by them. Then the inclusion given by $\langle \mathsf{LT}_\succ(m_1), \ldots, \mathsf{LT}_\succ(m_s) \rangle \subseteq \mathsf{LT}_\succ(M)$ holds.

Definition 19 Let $M \subseteq \mathrm{P}^r$ be a module. We denoted by $\mathsf{LT}_\succ\{M\}$ the *monomodule* $\{\mathsf{LT}_\succ(m) \mid m \in M \backslash \{0\}\} \subseteq \mathbb{T}^n \langle e_1, \ldots, e_r \rangle$.

Theorem 2 (Macaulay's Basis Theorem) *Let* $\mathrm{P} = K[x_1, \ldots, x_n]$ *be a polynomial ring over* K. *Let* $M \subseteq \mathrm{P}^r$ *be a* P-*submodule,* $\succ$ *a term ordering on* $\mathbb{T}^n \langle e_1, \ldots, e_r \rangle$, *and* B *the set of all terms in* $\mathbb{T}^n \langle e_1, \ldots, e_r \rangle \backslash \mathsf{LT}_\succ\{M\}$. *Then the residue classes of the elements of* B *form a basis of the* K-*vector space* P^r/M.

Proof See, for example, Sect. 1.5 in [14].

This theorem shows how to compute effectively P^r/M. First we need to compute $\mathsf{LT}_\succ(M)$ for a term ordering $\succ$, and then we represent each element as a unique and finite linear combination of the residue classes of the elements of $B = \mathbb{T}^n \langle e_1, \ldots, e_r \rangle \backslash \mathsf{LT}_\succ\{M\}$.

We have not seen yet how to compute efficiently $\mathsf{LT}_\succ(M)$. However, this is one of the subjects of the Sect. 2.2. Next, we recall the notion of homogeneous ideals and graded submodules.

Definition 20 The ring R is said to be a Γ-*graded ring* if there exist abelian subgroups R_γ, and

1. $R = \bigoplus_{\gamma \in \Gamma} R_\gamma$, and
2. for all $\gamma, \gamma' \in \Gamma$ it holds that $R_\gamma R_{\gamma'} \subseteq R_{\gamma+\gamma'}$.

We say that elements $r \in R_\gamma$ are *homogeneous of degree* γ and have all $\deg(r) = \gamma$. Additionally, the element 0 is a homogeneous element of R of every degree. Lastly, the decomposition of every element into its homogeneous components is unique. That property is a consequence of the direct sum in the Definition 20.

A natural way to extend the Definition 20 to R-modules is to use again the monoid Γ as the set of possible degrees. However doing so is not sufficiently general. The following definition underlines for now how to proceed.

Definition 21 Let R be a Γ-graded ring and $(\Omega, *)$ a Γ-monomodule. We say that M is a *Ω-graded R-module* if there exist abelian subgroups M_ω such that

1. $M = \bigoplus_{\omega \in \Omega} M_\omega$, and
2. for all $\gamma \in \Gamma$, $\omega \in \Omega$ it holds that $R_\gamma M_\omega \subseteq M_{\gamma * \omega}$.

Additionally, we say that a R-submodule N of M is a *Ω-graded R-submodule* of M if $N = \bigoplus_{\omega \in \Omega} (N \cap M_\omega)$, and that a Γ-graded submodule of R is a Γ-homogeneous ideal of R. For the sake of simplicity we call it *homogeneous ideal* of R if Γ is obvious from the context.

The following proposition and its corollary are very handy, especially for practical purposes. They allow us to quickly prove that some submodule is Ω-graded by exhibiting a homogeneous system of generators, and at the same time it give us a convenient representation of arbitrary homogeneous elements in terms of those homogeneous generators.

Proposition 3 *Let $N \subseteq M$ be a R-submodule and $N_\omega = N \cap M_\omega$ for all $\omega \in \Omega$. Then the following three conditions are equivalent.*

1. $N = \bigoplus_{\omega \in \Omega} N_\omega$.
2. *Let $n = \sum_{\omega \in \Omega} n_\omega$ be the decomposition of $n \in N$ into its homogeneous components, then $n_\omega \in N$ for all $\omega \in \Omega$.*
3. *There is a system of generators of N which consists only of homogeneous elements.*

Proof See, for example, Sect. 1.7 in [14].

Corollary 3 *Let $N \subseteq M$ be a Ω-graded R-submodule and $\{n_\gamma \mid \gamma \in \Gamma\}$ be a set of homogeneous generators of N. Suppose that the right-cancellation law holds in Ω and that $\omega \in \Omega$. Then we can say that every element $n \in N_\omega$ has a representation $n = \sum_{\gamma \in \Gamma} r_\gamma n_\gamma$ such that $r_\gamma \in R$ are homogeneous elements and* $\deg(r_\gamma) * \deg(n_\gamma) = \omega$ *for every $\gamma \in \Gamma$.*

Proof See, for example, Sect. 1.7 in [14].

Next we give a definition of homogenization and dehomogenization of polynomials.

Definition 22 Given any polynomial $p \in \mathrm{P} = R[x_1, \ldots, x_n]$ and an extra indeterminate x_0, we say that $p^h = x_0^{\deg(p)} p\left(\frac{x_1}{x_0}, \ldots, \frac{x_n}{x_0}\right) \in \mathrm{P}$ denotes the homogenization of p with respect to x_0. Hence, p^h is homogeneous polynomial of degree $\deg(p)$.

Definition 23 Conversely, for every homogeneous polynomial $\mathrm{P} \in R[x_0, \ldots, x_n]$ there exists a dehomogenization with respect to x_0 defined by $\mathrm{P}^{deh} = \mathrm{P}(1, x_1, \ldots, x_n) \in R[x_1, \ldots, x_n]$.

Let us recall now the definition of Hilbert functions and Hilbert series. These are two concepts that have an essential role throughout the next section.

Let $P = K[x_0, \ldots, x_n]$ be a graded polynomial ring, where each x_i is a homogeneous component of degree one.

Definition 24 The *Hilbert function* of the finitely generated P-module M is defined by the following map

$$\begin{aligned} \mathsf{HF}(M, d) : \mathbb{Z} &\longrightarrow \mathbb{Z} \\ v &\mapsto \dim(M_d) \end{aligned}$$

where M_d is the degree d part of M.

Definition 25 The *Hilbert polynomial* $q(d)$ is the unique polynomial in d such that the equivalence $q(d) = \mathsf{HF}(M, d)$ holds for all $d >> 0$. Let S be the support of M. Then, the Hilbert polynomial $q(d)$ has the following form $q(d) = \frac{i}{r!}d^r + \cdots$ where i is the degree of S and r is its dimension.

A compact representation of the dimensions of the homogeneous components is given by introducing the Hilbert series. This representation of the Hilbert function is desirable if we want to manipulate it on a computer. From now we shall use "computing the Hilbert function" and "computing the Hilbert series" interchangeably.

Definition 26 Let $M = \oplus_{n \geq 0} M_n$ be a finitely generated P-module M. The *Hilbert series* of M is defined by the generating function

$$\mathsf{HS}(M, t) = \sum_{d=-\infty}^{\infty} \mathsf{HF}(M, d) t^d$$

where $g(t)$ is a Laurent polynomial in t.

We conclude this introduction with a noteworthy theorem, which claims that, for a polynomial ring P and a homogeneous ideal I, the Hilbert series of P/I has always the form of a rational function.

Theorem 3 (Hilbert, Serre) *Given a graded polynomial ring* $\mathrm{P} = K[x_0, \ldots, x_n]$ *and a graded ideal* I *in* P, *then the Hilbert series of* P/I *can be expressed as*

$$\mathsf{HS}(P/I, t) = \frac{g(t)}{(1-t)^{n+1}}$$

where $g(t)$ *is a polynomial in* t *with integer coefficients. Moreover, the following relation holds,*

$$\mathsf{HS}(P/I, t) = \mathsf{HS}(P, t) - \mathsf{HS}(I, t)$$

Proof See, for example, Sect. 5.2 in [15].

2.2 Gröbner Bases Theory

Let us assume for now that we want to compute $\mathsf{LT}_{\succ}(I)$ for a given ideal $I = \langle f_1, \ldots, f_s\rangle$. It is worth mentioning that a direct computation of $\mathsf{LT}_{\succ}(I)$ is impossible because it requires the computation of the leading term of each polynomial in the ideal I. As an alternative, we might want to consider the monomial ideal generated by the leading terms of the generators,

$$\langle \mathsf{LT}_{\succ}(f_1), \ldots, \mathsf{LT}_{\succ}(f_s)\rangle \subseteq \mathsf{LT}_{\succ}(I)$$

At this point, one might ask: is it is possible to produce a set of generators for which these two ideals are the same? The answer to this question is exactly the notion of a Gröbner basis.

Definition 27 Given an ideal $I \subseteq \mathrm{P}$ and a fixed term ordering $\succ$. A finite set of polynomials $\mathscr{G} \subseteq I\backslash\{0\}$ is a *Gröbner basis* of $I\backslash\{0\}$ if and only if the leading term of every element of I is divisible by one term in the set $\mathscr{G}$,

$$\langle \mathsf{LT}_{\succ}(g) \mid g \in \mathscr{G}\rangle = \mathsf{LT}_{\succ}(I)$$

Henceforth we have all the tools that are necessary to check if a set of polynomials is a Gröbner basis for some ideal I. Let us see an example of how to do it:

Example 1 Let $\succ_{\mathrm{Lex}}$ be a fixed term ordering and $I = \langle f_1, f_2\rangle = \langle x^3 - 2xy, x^2y - 2y^2 - x\rangle$ an ideal in P. Then we have that $\langle f_1, f_2\rangle$ is not a Gröbner basis for I. Although we have $x^2 = yf_1 - xf_2 \Rightarrow x^2 \in I \Rightarrow x^2 = \mathsf{LT}_{\succ}(x^2) \in \langle \mathsf{LT}_{\succ}(I)\rangle$, we also have that x^2 is not divisible by $\mathsf{LT}_{\succ}(f_1) = x^3$ or by $\mathsf{LT}_{\succ}(f_2) = x^2y$, which means that $x^2 \notin \langle \mathsf{LT}_{\succ}(f_1), \mathsf{LT}_{\succ}(f_2)\rangle$.

Up to now, we saw that Gröbner bases are not defined uniquely, e.g, the Definition 27 does not require a Gröbner basis to be minimal. Therefore, one can add redundant elements of I to $\mathcal{G}$ and get another Gröbner basis. This means that any finite subset of I that contains a Gröbner basis for I is also a Gröbner basis. Nevertheless, this lack of uniqueness can be solved by readjusting the definition of Gröbner basis into something called reduced Gröbner bases.

Definition 28 A Gröbner basis $\mathcal{G}$ is a *reduced Gröbner basis* if

1. for each $g \in \mathcal{G}$, the coefficient $\mathsf{LC}_{\succ}(g)$ is 1,
2. $\mathsf{LT}_{\succ}(I)$ is minimally generated by the set $\{\mathsf{LT}_{\succ}(g) : g \in \mathcal{G}\}$, and
3. no trailing term of any $g \in \mathcal{G}$ lies in $\mathsf{LT}_{\succ}(I)$.

A basic concept behind the theory of Gröbner bases is the idea of *reduction* and *normal form*.

Definition 29 Let f be a polynomial in P and $\mathcal{G}$ a Gröbner basis for an ideal $I \subseteq$ P. Then there is a unique $r \in$ P such that:

- the polynomial $f - r$ is in I, and
- no term of r is divisible by any term in $\mathsf{LT}_{\succ}(\mathcal{G})$.

The polynomial r is called the *normal form* of f with respect to $\mathcal{G}$, and can be obtained by reducing f repeatedly by polynomials of $\mathcal{G}$, until it is *irreducible*. We denote the normal form of f as $r = \mathsf{NR}_{\succ,\mathcal{G}}(f)$.

Theorem 4 *$\mathcal{G}$ is a Gröbner basis of an ideal I if and only if the following holds*

$$f \in I \Longleftrightarrow \mathsf{NR}_{\succ,\mathcal{G}}(f) = 0$$

Proof Let f be an arbitrary polynomial in the ideal I. Then the division of f by $\mathcal{G}$ yields the following form

$$f = e_1 g_1 + \cdots + e_m g_m + r$$

Hence $r \in I$ because $f - r \in I$. Assume that $r \neq 0$. Then we can find a k such that $\mathsf{LT}_{\succ}(g_k)$ divides $\mathsf{LT}_{\succ}(r)$, since $\mathcal{G}$ is a Gröbner basis. This is a contradiction to the fact that r is reduced with respect to $\mathcal{G}$. Thus, r must be equal to zero.

Another worth mentioning theorem that shall be introduced concerns the existence of reduced Gröbner bases.

Theorem 5 *Every nonzero polynomial ideal has a unique reduced Gröbner basis.*

Proof See, for example, the proof of Theorem 2.4.13 of Sect. 2.4 in [14].

2.2.1 Buchberger's Algorithm

Buchberger introduced in his PhD dissertation [3] an algorithm to systematically compute a reduced Gröbner basis $\mathscr{G}$ for any ideal I. It turns out that the concept of a Gröbner bases is intimately related to the concept of S-polynomials.

Definition 30 Let $g_1, g_2 \in \mathrm{P}$. The *S-polynomial* of g_1 and g_2 is defined to be

$$S_{1,2} = \frac{p}{\mathsf{LM}_{\succ}(g_1)} g_1 - \frac{p}{\mathsf{LM}_{\succ}(g_2)} g_2$$

where p is the least common multiple: $p = \mathsf{lcm}(\mathsf{LT}_{\succ}(g_1), \mathsf{LT}_{\succ}(g_2))$,

Although the notion of S-polynomials is the nucleus of *algorithmic* Gröbner bases theory, the notion of Gröbner bases is independent of the notion of S-polynomials as it also gives many interesting results for the *non-algorithmic* polynomial ideal theory.

Theorem 6 (Buchberger's Criterion) *Let* $I = \langle g_1, \ldots, g_m \rangle \subseteq \mathrm{P}$, $\mathscr{G}$ *a set of polynomials* $\{g_1, \ldots, g_m\}$, *and* $\mathbb{B} = \{(i, j) \mid 1 \leq i < j \leq m, e_i = e_j\}$, *where* $\mathsf{LM}_{\succ}(g_i) = c_i t_i e_i$. *Then the following conditions are equivalent:*

1. *The set* $\mathscr{G}$ *is a Gröbner basis of* I *w.r.t.* $\succ$.
2. *For all pairs* $(i, j) \in \mathbb{B}$, *we have* $\mathrm{NR}_{\succ,\mathscr{G}}(\mathrm{S_{i,j}}) = 0$.

Proof See, for example, Sect. 2.5 in [16].

With this criterion, one can easily recreate Buchberger's algorithm [1985], which computes the reduced Gröbner basis $\mathscr{G}$ for any finite set of polynomials. The basic Buchberger's algorithm is the following:

Input:
  A set F of generators $\{f_1, \ldots, f_m\}$ of the ideal I.
  A term ordering $\succ$.
Output:
  A Gröbner basis of I.
Set $\mathscr{G}$ equal to F;
Let B be the set of pairs $\{(i, j) \mid g_i, g_j \in \mathscr{G}, i > j\}$;
while $B \neq \emptyset$ **do**
  Choose a pair $(i, j) \in B$;
  Set $B := B \setminus \{(i, j)\}$;
  Compute S_{ij} and let $h = \mathrm{NR}_{\succ,\mathscr{G}}(\mathrm{S_{ij}})$;
  if $h = 0$ **then**
    Go to line 3;
  end
  if $h \neq 0$ **then**
    Add h to $\mathscr{G}$;
    Build new all critical pairs containing h and add them to B;
    Go to step 3;
  end
end
return $\mathscr{G}$

Algorithm 1: Buchberger's Algorithm

Although, Buchberger's algorithm is undeniably simple, the proof of its termination is not trivial. The algorithm initializes with a finite number of pairs of polynomials, which correspond to the S-polynomials to be computed. Some of those pairs lead to zero reductions while others contribute to the Gröbner basis. Hence, $\mathscr{G}$ is expected to grow, as well as the number of additional pairs that have to be considered.

Theorem 7 *Buchberger's algorithm terminates after a finite number of steps.*

Proof Let $\mathscr{G}_n$ denote the trial Gröbner basis that is produced by the algorithm at the n-th iteration step. We know that $\mathscr{G}_1 \subset \mathscr{G}_2 \subset \mathscr{G}_3 \subset \ldots$, however, the most important part is that the sets of leading terms increase as well: $\mathsf{LT}_{\succ}(\mathscr{G}_1) \subset \mathsf{LT}_{\succ}(\mathscr{G}_2) \subset \mathsf{LT}_{\succ}(\mathscr{G}_3) \subset \cdots$. By Dickson's Lemma [1913], this process ends in a finite number of steps, once the sequence of leading terms stabilizes. See [3] for a complete proof.

2.2.2 Introduction to Signature-Based Strategies

The main idea of the signature-based strategies for computing a Gröbner basis is to associate each polynomial with a signature on which the criteria and reduction steps depend. The concept was initially proposed by [7] together with his F5 algorithm [7], and has become extremely popular partly due to its good performances.

The most efficient variants of F5 are: the F5C algorithm by [5], the F5 with extended criteria by [11], the Gao-Guan-Volny (G2V) [8], and the Gao-Volny-Wang (GVW) [9]. The main differences between these algorithms are the extra conditions implemented to ensure correctness of the strategy.

Suppose, for example, that f and g are two polynomials, where t_f and t_g are terms such that the leading terms of $t_f f$ and $t_g g$ are the same. Then a necessary condition to reject the critical pair of f and g is that, there exists some known polynomial h such that its signature is a factor of $t_f f'$s or $t_g g'$s signature. We shall see that this condition alone is not sufficient to ensure correctness. Thus, existing implementations use different extra conditions.

In this section, we give a brief introduction to signature-based strategies. First, we recall the concept of polynomial signature. Then, we describe a generic footprint for signature-based algorithms. Lastly, we introduce a detailed description of the GVW algorithm. For proofs of the correctness and termination, see for example, [4, 5, 7, 8, 11, 17]. However, most of these proofs assume the input to be homogeneous, or the critical pair that is associated to the smallest signature to be computed first. These are extra requirements can affect the performance of the algorithm by making it less efficient.

Let $\mathrm{P} = K[x_1, \ldots, x_n]$ be a polynomial ring over a field K and $\succ$ be a fixed term ordering in P. Let $I = \langle F \rangle \subset R$ be an ideal generated by the elements of $F = \{g_1, \ldots, g_m\}$, where each $g_i \in \mathrm{P}$.

$$I = \langle g_1, \ldots, g_m \rangle = \{u_1 g_1 + \cdots + u_m g_m : u_1, \ldots, u_m \in \mathrm{P}\} \subseteq \mathrm{P} \tag{2}$$

Then we can define the following map

$$\phi : \mathrm{P}^m \longrightarrow I$$
$$\sum_{i=1}^{m} u_i \mathbf{e}_i \mapsto \sum_{i=1}^{m} u_i g_i$$

such that the $u_i{}'$s are polynomials in P, and $\mathbf{e}_1, \ldots, \mathbf{e}_m$ are the canonical generators of the free P-module P^m. We have that the elements $\mathbf{u} \in R^m$ with $\phi(\mathbf{u}) = 0$ are called *syzygies* of $g_1, \ldots, g_m$. We use the notation $\mathbf{H} = \mathrm{Syz}(F)$ to represent the module of all such syzygies.

$$\mathbf{H} = \{(u_1, \ldots, u_m) \in P^m : u_1 g_1 + \cdots + u_m g_m = 0\} \tag{3}$$

We denote elements of P^m, which we see as row vectors, with bold letters e.g. $\mathbf{g}$, $\mathbf{u}$ etc. Next, we consider the following P-submodule of $\mathrm{P}^m \times \mathrm{P}$:

$$M = \{(\mathbf{u}, v) \in \mathrm{P}^m \times \mathrm{P} : \mathbf{u}\mathbf{g}^t = v\} \tag{4}$$

generated by

$$(\mathbf{e}_1, g_1), (\mathbf{e}_2, g_2), \ldots, (\mathbf{e}_m, g_m) \tag{5}$$

Now that all the settings are in place, we introduce our first definition of polynomial signature.

Definition 31 [7] Let $\mathrm{P} = K[x_1, \ldots, x_n]$ and v a polynomial in the ideal $I \subseteq \mathrm{P}$. Let $u_1, \ldots, u_m \in \mathrm{P}$ and $\mathbf{u} \in \mathrm{P}^m$ such that $v = \phi(\mathbf{u})$, where $\mathbf{u} := \sum_{i=1}^{m} u_i \mathbf{e}_i$. We say that $\mathsf{LT}_{\succ}(\mathbf{u})$ is a *signature* of $(\mathbf{u}, v) \in \mathrm{P}^m \times \mathrm{P}$. Moreover, $v \in \mathrm{P}$ has an unique and minimal signature if a term ordering is fixed.

Signature-based algorithms follow also Buchbergers algorithm footprints. Hence, we shall define a similar concept to the one of critical pairs.

Definition 32 Suppose $(\mathbf{u}_1, v_1), (\mathbf{u}_2, v_2) \in \mathrm{P}^m \times \mathrm{P}$ are two pairs with v_1 and v_2 both nonzero. Let

$$t_1 = \frac{\mathsf{lcm}(\mathsf{LT}_{\succ}(v_1), \mathsf{LT}_{\succ}(v_2))}{\mathsf{LT}_{\succ}(v_1)}, \quad t_2 = \frac{\mathsf{lcm}(\mathsf{LT}_{\succ}(v_1), \mathsf{LT}_{\succ}(v_2))}{\mathsf{LT}_{\succ}(v_2)}$$

Suppose $\max(t_1 \mathsf{LT}_{\succ}(\mathbf{u}_1), t_2 \mathsf{LT}_{\succ}(\mathbf{u}_2)) = t_i \mathsf{LT}_{\succ}(\mathbf{u}_i)$. Then

- We call $t_i(\mathbf{u}_i, v_i) = (t_i\mathbf{u}_i, t_i v_i)$ a *J-pair*[1] of $(\mathbf{u}_1, v_1)$ and $(\mathbf{u}_2, v_2)$, and
- $t_i \mathsf{LT}_{\succ}(\mathbf{u}_i)$ their *J-signature*.

Since the J-signature of $(\mathbf{u}_1, v_1)$ and $(\mathbf{u}_2, v_2)$ is unique, then we are allowed to pick either $t_1(\mathbf{u}_1, v_1)$ or $t_2(\mathbf{u}_2, v_2)$ as a J-pair if $t_1 \mathsf{LT}_{\succ}(\mathbf{u}_1) = t_2 \mathsf{LT}_{\succ}(\mathbf{u}_2)$. Also, it is only necessary to keep one J-pair per distinct J-signature.

[1] J means "joint" of the two pairs.

The terms t_1 and t_2 used in J-pairs are exactly the same as those used in Buchberger's S-polynomial $t_1 v_I - ct_2 v_2$, where $c = \mathsf{LC}_\succ(v_I)/\mathsf{LC}_\succ(v_2)$. The main difference between these two concepts is that J-pairs postpone the computation of S-polynomials until the reduction phase, as we work with their signatures. This idea should become more clear with the introduction of the GVW algorithm.

We revise now the reduction process, which shall depend as well on the concept of signature.

Definition 33 Let $(\mathbf{u}_1, v_1), (\mathbf{u}_2, v_2) \in \mathrm{P}^m \times \mathrm{P}$ be any two pairs. If v_2 is nonzero, $\mathsf{LT}_\succ(v_2)$ divides $\mathsf{LT}_\succ(v_I)$ and $\mathsf{LT}_\succ(t\mathbf{u}_2) \succeq \mathsf{LT}_\succ(\mathbf{u}_1)$ then we say $(\mathbf{u}_1, v_I)$ is *top-reducible* by $(\mathbf{u}_2, v_2)$. The corresponding *top-reduction* is

$$(\mathbf{u}_1, v_I) - ct(\mathbf{u}_2, v_2) = (\mathbf{u}_I - ct\mathbf{u}_2, v_I - ctv_2), \tag{6}$$

where $t = \mathsf{LT}_\succ(v_1)/\mathsf{LT}_\succ(v_2)$ and $c = \mathsf{LC}_\succ(v_1)/\mathsf{LC}_\succ(v_2)$.
When v_2 is zero and $\mathbf{u}_1, \mathbf{u}_2 \neq 0$, then we say that $(\mathbf{u}_1, v_1)$ is top-reducible by $(\mathbf{u}_2, 0)$ if $\mathsf{LT}_\succ(\mathbf{u}_2)$ divides $\mathsf{LT}_\succ(\mathbf{u}_1)$. For such scenario the top-reduction is

$$(\mathbf{u}_1, v_I) - ct(\mathbf{u}_2, v_2) = \left(\mathbf{u}_I - \frac{\mathsf{LM}_\succ(\mathbf{u}_I)}{\mathsf{LM}_\succ(\mathbf{u}_2)}\mathbf{u}_2, v_I\right)$$

Every time we perform a top-reduction, we cancel the leading term in the v-part without increasing the signature of $(\mathbf{u}_1, v_1)$. Moreover, we can say that,

Definition 34 A top-reduction is called *regular top-reduction*, if

$$\mathsf{LT}_\succ(\mathbf{u}_1 - ct\mathbf{u}_2) = \mathsf{LT}_\succ(\mathbf{u}_1),$$

and *super top-reduction* otherwise. This means that the signature of $(\mathbf{u}_1, v_1)$ becomes smaller under a super top-reduction and remains unchanged under a regular top-reduction. Observe that a super top-reduction happens if

$$\mathsf{LT}_\succ(t\mathbf{u}_2) = \mathsf{LT}_\succ(\mathbf{u}_1) \text{ and } \frac{\mathsf{LC}_\succ(\mathbf{u}_1)}{\mathsf{LC}_\succ(\mathbf{u}_2)} = \frac{\mathsf{LC}_\succ(\mathrm{v}_1)}{\mathsf{LC}_\succ(\mathrm{v}_2)}$$

Furthermore, if a pair $(\mathbf{u}_1, v_1)$ is top-reducible by $(\mathbf{u}_2, 0)$, then the top-reduction is always a super top-reduction. Hence we do not have to perform the reduction. Also, a pair $(\mathbf{u}_1, 0)$ is never top-reducible by $(\mathbf{u}_2, v_2)$ for $v_2 \neq 0$. Note that we do not have to perform reductions for the super top-reduction case.

Lemma 1 *Let t be a term in P. If a pair $t(\mathbf{u}_1, v_1)$ is (regular) top-reducible by $(\mathbf{u}_2, v_2)$, where both v_1 and v_2 are nonzero, then $t_1(\mathbf{u}_1, v_1)$ is a J-pair of $(\mathbf{u}_1, v_1)$ and $(\mathbf{u}_2, v_2)$, where*

$$t_1 = \frac{\mathsf{lcm}(\mathsf{LT}_\succ(v_1), \mathsf{LT}_\succ(v_2))}{\mathsf{LT}_\succ(v_1)}$$

and t_1 is a divisor of t. Furthermore, $t_1(\mathbf{u}_1, v_1)$ is (regular) top-reducible by $(\mathbf{u}_2, v_2)$.

Proof See, for example, [8].

Proposition 4 *Suppose that* $L = \{(\mathbf{u}_1, v_1), \ldots, (\mathbf{u}_k, v_k)\}$ *is a strong Gröbner basis for* M*, that is, every pair* $(\mathbf{u}, v) \in M$ *is top-reducible by some pair in* L*. Then one can derive the following conclusions*

- *The set* $\mathcal{G}_0 = \{v_i : 1 \leq i \leq k\}$ *is a Gröbner basis for the ideal* $I = \langle g_1, \ldots, g_m \rangle$*, and*
- *The set* $\mathcal{G}_1 = \{\mathbf{u}_i : v_i = 0,\ 1 \leq i \leq k\}$ *is a Gröbner basis for the syzygy module of* $\mathbf{g} = (g_1, \ldots, g_m)$*.*

A strong Gröbner basis for $M \subset \mathrm{P}^m \times \mathrm{P}$ is a Gröbner basis for M as a submodule of P^{m+1}. However the converse may not be true for an arbitrary submodule M of P^{m+1}. This is why we call the basis a strong Gröbner basis.

Proof In order to prove that $\mathcal{G}_1$ is a Gröbner basis for the syzygy module of $\mathbf{g}$, one can deduce from equations (3) and (4) that we must have $(\mathbf{u}, 0) \in M$ for any $\mathbf{u} = (u_1, \ldots, u_m)$ in the syzygy module of $\mathbf{g}$. Now, suppose that $(\mathbf{u}, 0)$ is top-reducible by some pair $(\mathbf{u}_i, v_i)$ in L. Then by Definition (34), v_i must be zero. Therefore, $\mathbf{u}_i \in \mathcal{G}_1$ and $\mathsf{LT}_{\succ}(\mathbf{u})$ is reducible by $\mathsf{LT}_{\succ}(\mathbf{u}_i)$.

Now, to prove that $\mathcal{G}_0$ is a Gröbner basis for I, assume $v \in I$ and nonzero. Then we know there exists $\mathbf{u} = (u_1, \ldots, u_m) \in \mathrm{P}^m$ so that $\mathbf{u}\mathbf{g}^t = v$, hence $(\mathbf{u}, v) \in M$. Among all such $\mathbf{u}$, we pick the one having the smallest leading term. Since $(\mathbf{u}, v) \in M$, it is top-reducible by some $(\mathbf{u}_i, v_i)$ where $1 \leq i \leq k$. Assume, by absurd, that $v_i = 0$. Then we could use $(\mathbf{u}_i, 0)$ to reduce $(\mathbf{u}, v)$ to get a $\mathbf{u}'$ such that $\mathbf{u}'\mathbf{g}^t = v$, with $\mathsf{LT}_{\succ}(\mathbf{u}')$ smaller than $\mathsf{LT}_{\succ}(\mathbf{u})$, which contradicts the initial minimality of $\mathsf{LT}_{\succ}(\mathbf{u})$. So $v_i \neq 0$ and $\mathsf{LT}_{\succ}(v_i)$ divides $\mathsf{LT}_{\succ}(v)$. Therefore, we have that $\mathcal{G}_0$ is a Gröbner basis for I.

Definition 35 Let L be any set of pairs in M as defined in (4). We say that $(\mathbf{u}, v)$ is *eventually super top-reducible* by L if there is a sequence of regular top-reductions of $(\mathbf{u}, v)$ by pairs in L that reduce $(\mathbf{u}, v)$ to a pair $(\mathbf{u}', v')$ that is no longer regular top-reducible by L but is super top-reducible by at least one pair in L.

Theorem 8 *Let* L *be any set of pairs in* M *as defined in* (4)*. Then* L *is a strong Gröbner basis for* M *if and only if for every distinct J-signature from* L *there is at least one J-pair from* L *with the same J-signature that is eventually super top-reducible by* L*.*

Proof See [8] for a proof of this theorem.

Theorem 9 *If the Algorithm 2 terminates, then* V *is a Gröbner basis for* $I = \langle g_1, g_2, \ldots, g_m \rangle$ *and* $\mathbf{H}$ *is a Gröbner basis for the leading terms of the syzygy module of* $(g_1, g_2, \ldots, g_m)$*.*

Proof See [8] for a proof of this theorem.

The algorithm that puts together all the pieces follows next.

Input:
List of polynomials $g_1, \dots, g_m \in \mathrm{P} = K[x_1, \dots, x_n]$,
A term ordering for P, and
A term ordering on P^m
Output:
A Gröbner basis V of $I = \langle g_1, \dots, g_m \rangle$, and
A Gröbner basis H of $\mathrm{LT}_\succ(\mathbf{H})$, the leading terms of the syzygy module
Variables:
U a list of terms T_i, representing signatures of $(\mathbf{u}_i, v_i) \in M$,
V a list of polynomials v_i for $(\mathbf{u}_i, v_i) \in M$,
H a list of $\mathrm{LT}_\succ(\mathbf{u})$ were $\mathbf{u} \in R^m$ is a syzygy found so far,
JP a list of pairs (t, i) s.t. $t(u_i, v_i)$ is the J-pair of $(\mathbf{u}_i, v_i)$ and $(\mathbf{u}_j, v_j)$ for some $j \neq i$.
Algorithm:
Initialize U as $\{\mathbf{e}_1, \dots, \mathbf{e}_m\}$;
Initialize V as $\{g_1, \dots, g_m\}$;
Add the leading terms of the principle syzygies $g_j \mathbf{e}_i - g_i E_j$ for $1 \leq i < j \leq m$ to H ;
Compute all the J-pairs of $(\mathbf{e}_1, g_1), \dots, (\mathbf{e}_m, g_m)$;
Add to the list JP all J-pairs whose signatures are distinct and not reducible by H ;
while *JP is not empty* **do**
  Take a minimal pair (t, i) from JP (w.r.t. signature) ;
  Delete the pair (t, i) from the list JP ;
  while $t(T_i, v_i)$ *is regular top-reducible by the pairs in* (U, V) **do**
    Perform regular top-reduction, say to get(T, v) ;
  end
  if $v = 0$ **then**
    Append T to H ;
    Delete every J-pair (t, j) in JP whose signature tT_j is divisible by T ;
  end
  if $v \neq 0$ *and* (T, v) *is not super top-reducible by* (U, V) **then**
    Append T to U and v to V;
    Form all J-pairs for (T, v) and (T_j, v_j), $1 \leq j \leq |U| - 1$,and ;
    Consider only J-pairs whose signatures are not reducible by H, and ;
    Add to JP only the J-pair with minimal $\mathrm{LT}_\succ(v)$ for each distinct signature T ;
    Add leading terms of the principle syzygies, $vT_j - v_j T$ for $1 \leq j \leq |U| - 1$, to H ;
  end
end
return V and H;

Algorithm 2: The GVW algorithm

3 The Hilbert-Driven Strategy

In the previous section, we introduced the notion of Hilbert series and we have shown its connection to the Hilbert polynomial. In this section, we introduce a new criterion to signature-based strategies that is based on the so-called *Hilbert-driven Gröbner basis algorithm*, originally presented by [18].

Current attempts based on this strategy have a common drawback: one needs to know the Hilbert series beforehand in order to take advantage of it. Some work about how to compute it efficiently can be found in [2]. In some special cases we have all the information we need about the Hilbert series without any further computations.

Theorem 10 *Let $\succ$ be a term ordering on P, and $I \subset P$ a homogeneous ideal. Then we have the following relation*

$$\mathsf{HP}(\mathrm{P}/I, t) = \mathsf{HP}(\mathrm{P}/\,\mathsf{LT}_\succ(I), t)$$

Proof See for example Sect. 5.2 in [10].

Using the above theorem one can derive a few other properties, which are nicely described through the following statement.

Corollary 4 *Let $I \in P$ be an ideal and $\succ$ a term ordering. Let $\mathscr{G} = \{g_1, \ldots, g_s\} \subset I$. Then it holds:*

1. $\mathsf{HF}(P/\mathsf{LT}_{\succ}(I), d) \leq \mathsf{HF}(P/\mathsf{LT}_{\succ}(\mathscr{G}), d)$ *for all* d.
2. *If* $\mathsf{HF}(P/\mathsf{LT}_{\succ}(I), d) = \mathsf{HF}(P/\mathsf{LT}_{\succ}(\mathscr{G}), d)$ *for all* d, *then* $\mathscr{G}$ *is a Gröbner basis for* I.

Proof The proof follows easily from the fact that $\mathsf{LT}_{\succ}(\mathscr{G}) \subset \mathsf{LT}_{\succ}(I)$. Having $\mathsf{LT}_{\succ}(\mathscr{G}) \subset \mathsf{LT}_{\succ}(I)$, the equality of the Hilbert functions follows from the equality of the leading ideals, i.e. $\mathsf{LT}_{\succ}(\mathscr{G}) = \mathsf{LT}_{\succ}(I)$. But this is just the definition of $\mathscr{G}$ being a Gröbner basis for I.

Corollary 5 *Let $\succ_1$ and $\succ_2$ be two term orderings on* P *and $I \subseteq$ P an ideal.*

1. *If I is homogeneous, then* $\mathsf{HF}(P/\mathsf{LT}_{\succ_1}(I), d) = \mathsf{HF}(P/I, d) = \mathsf{HF}(P/\mathsf{LT}_{\succ_2}(I), d)$ *for all* d.
2. *If I is inhomogeneous, then* $\mathsf{HF}(P/\mathsf{LT}_{\succ_1}(I), d) = \mathsf{HF}(P/I, d) - \mathsf{HF}(P/I, d-1) = \mathsf{HF}(P/\mathsf{LT}_{\succ_2}(\mathscr{G}), d)$ *for all* d.

Proof See for example [18].

Another important corollary, which is given by Traverso in [18], describes how to use the Hilbert series to improve the computations of Gröbner basis for inhomogenous ideals.

Corollary 6 *Let $I \subset$ P be an ideal and $\succ_1$ and $\succ_2$ two term orderings on* P. *Let $\mathscr{G}_1$ be a Gröbner basis for I w.r.t. $\succ_1$. Set $\mathscr{G}_1$ as an input for the computation of a Gröbner basis $\mathscr{G}_2$ for I w.r.t. $\succ_2$. Then we can use the following variant algorithm:*

1. *Consider critical pairs by increasing degree.*
2. *If the degree decreases during a reduction step, then the reduced element can be deleted and the next pair can be computed.*

This corollary is very useful when the computation of a Gröbner basis w.r.t. $\succ_1$ is easier than the computation w.r.t.$\succ_2$.

The next theorem from [18] gives one of the most important results of this section.

Theorem 11 *Let I and J be two homogeneous ideals in* P *such that $J \subset I$. By Theorem there exist polynomials $p(t) = \sum_{i=0}^{v} p_i t^i$ and $q(t) = \sum_{j=0}^{w} q_j t^j$ such that the corresponding Hilbert series are*

$$\mathsf{HS}(P/I, t) = \frac{p(t)}{(1-t)^n} \text{ and } \mathsf{HS}(P/J, t) = \frac{q(t)}{(1-t)^n}.$$

Then the following conditions are equivalent:

1. $\mathsf{HF}(\mathrm{P}/I, t) = \mathsf{HF}(\mathrm{P}/J, t)$ *for all* $1 \leq t \leq d-1$ *and* $\mathsf{HF}(\mathrm{P}/I, d) < \mathsf{HF}(\mathrm{P}/J, d)$.
2. $p(i) = q(i)$ *for* $1 \leq i \leq d-1$ *and* $p(d) < q(d)$.

Definition 36 Let $(p_1, \ldots, p_r)$ be a sequence of polynomials in P. Let F be a finitely generated module in M. We call $(p_1, \ldots, p_r)$ a *sequence regular* if for each $1 \leq i \leq r$ it holds that p_i is not a zero divisor in $F/\langle p_1, \ldots, p_{i-1}\rangle$

From the definition above, we have that an important property of a complete intersection $V(I)$ is that we can compute the corresponding Hilbert series of I without computing a Gröbner basis for I beforehand. The following lemma shows how.

Lemma 2 *If* $V(I)$ *is a complete intersection for* $I = \langle f_1, \ldots, f_r\rangle$ *where* f_i *is homogeneous of degree* $\deg(f_i) = d_i$ *for all* $1 \leq i \leq r$*, then the Hilbert series is given by*

$$\mathsf{HS}(\mathrm{P}/I, t) = \frac{\prod_{i=1}^{r}(1 - t^{d_i})}{(1-t)^n} \tag{7}$$

Now we have all the tools required to describe the idea behind the Hilbert-driven algorithm.

Let $\succ_1$ be a term ordering on P and let I be the ideal for which we want to compute a Gröbner basis. Assume that we know the Hilbert function $\mathsf{HF}(\mathrm{P}/I, t)$. Keep in mind that it can be a consequence of one of the following facts

1. A previous Gröbner basis computation for I w.r.t. some other term ordering.
2. Or a special shape of the ideal I: a complete intersection where the formula is known (see Eq. 7).

If I is homogeneous then we compute the Gröbner basis $\mathcal{G}$ for I by increasing degree. Assume that we have an intermediate Gröbner basis $\mathcal{G}$ for some degree $d \geq 0$. Then we compute the Hilbert function $\mathsf{HF}(\mathrm{P}/\mathsf{LT}_{\succ}(\mathcal{G}), t)$. It holds that $\mathcal{G}$ is a Gröbner basis for I if

$$\mathsf{HF}(\mathrm{P}/\mathsf{LT}_{\succ}(\mathcal{G}), t) = \mathsf{HF}(t) \text{ for all } t \leq d.$$

Next, we shall see how to incorporate this strategy into signature-based algorithms. Our implementation is based on the footprints of the algorithm of [9]. Thereby, the output of the algorithm not only returns a Gröbner basis of the ideal I, but also a Gröbner basis of the leading terms of the syzygy module of the same ideal.

The idea of Algorithm 3 is the following. Let α_k and β_k be two variables used to keep track of the number of generators of degree k in the Gröbner basis and in the leading terms of the syzygy module, which shall then used to avoid unnecessary computations of the Hilbert series. Then, if the equality $\mathsf{HF}(\mathrm{P}/\mathsf{LT}_{\succ}(\mathcal{G}), k) = \mathsf{HF}(k) - \mathsf{HF}(\mathrm{P}/H, k-d)$ holds for all k, we have that $\mathcal{G}$ is a Gröbner basis of I and H is a Gröbner basis for the leading terms of the syzygy module. See lines 28–29. Otherwise, we know that there exists a minimal degree d for each we need to compute more generators. That degree d also gives us a bound for the minimal degree of the S-polynomials that we need to compute.

Next we give the pseudo-code of the algorithm that puts together all the ideas discussed. The reader should keep in mind that this strategy highly depends on the order in which the J-pairs are processed. If the first J-pairs are the useful ones, the optimization is best. If those are at the end of the list of pairs to be reduced, then we still have to compute the zero reductions of all useless pairs before them.

Input:
An ideal $I = \langle g_1, \dots, g_m \rangle \in \mathrm{P} = K[x_1, \dots, x_n]$,
A term ordering for P, and a term ordering on P^m
$\mathsf{HF}(w) := \mathsf{HF}(\mathrm{P}/I, w)$ **the Hilbert function of** I **(Optional)**
Output:
A Gröbner basis V for I and,
A Gröbner basis H for $\mathsf{LT}_{\succ}(\mathbf{H})$, the leading terms of the syzygy module
Variables:
U a list of terms T_i, representing sig. of $(\mathbf{u}_i, v_i) \in M$,
V a list of polynomials v_i for $(\mathbf{u}_i, v_i) \in M$,
H a list of $\mathsf{LT}_{\succ}(\mathbf{u})$ were $\mathbf{u} \in \mathrm{P}^m$ is a syzygy found so far,
JP a list of pairs (t, i) s.t. $t(u_i, v_i)$ is the J-pair of $(\mathbf{u}_i, v_i)$ and $(\mathbf{u}_j, v_j)$ for some $j \neq i$.
Algorithm:
U = $\{\mathrm{e}_1, \dots, \mathrm{e}_m\}$ and V = $\{g_1, \dots, g_m\}$;
Add the leading terms of the principle syzygies $g_j \mathrm{e}_i - g_i \mathrm{e}_j$ for $1 \leq i < j \leq m$ to H ;
Compute all the J-pairs of $(\mathrm{e}_1, g_1), \dots, (\mathrm{e}_m, g_m)$;
Add to the list JP all J-pairs whose signatures are distinct and not reducible by H ;
Initialize d **with the minimal degree in JP** ;
while JP is not empty **do**
  Take a minimal pair (t, i) from JP (with respect to signature) ;
  Delete the pair (t, i) from the list JP ;
  while *$t(T_i, v_i)$ is regular top-reducible by the pairs in (U, V)* **do**
    Perform regular top-reductions if the degree is $\geq d$**, say to get** (T, v) **of degree** k**. Else Goto 6** ;
  end
  if $v = 0$ **then**
    Append T to H ;
    Delete every J-pair (t, j) in JP whose signature tT_j is divisible by T ;
    $\beta_k \leftarrow \beta_k + 1$;
  end
  if *$v \neq 0$ and (T, v) is not super top-reducible by (U, V)* **then**
    Append T to U and v to V;
    Form all J-pairs for (T, v) and (T_j, v_j), $1 \leq j \leq |U| - 1$, and ;
    Consider only J-pairs whose signatures are not reducible by H, and ;
    Add to JP J-pairs with minimal $\mathsf{LT}_{\succ}(v)$ for each distinct signature T ;
    Add LTs of the principle syzygies, $vT_j - v_j T$ for $1 \leq j \leq |U| - 1$, to H ;
    Update β_k ;
    $\alpha_k \leftarrow \alpha_k + 1$;
  end
  if $\mathsf{HF}(\mathrm{P}/\mathsf{LT}_{\succ}(V), w) = \mathsf{HF}(w) - \mathsf{HF}(\mathrm{P}/\mathrm{H}, \mathrm{w} - \mathrm{d})$ **for all** w **then**
    return V and H;
  end
  $d \leftarrow \min\{w \in \mathbb{N} \mid \mathsf{HF}(\mathrm{P}/\mathsf{LT}_{\succ}(V), w) \neq \mathsf{HF}(w) - \mathsf{HF}(\mathrm{P}/\mathrm{H}, \mathrm{w} - \mathrm{d})\}$;
end
return V and H;

Algorithm 3: Signature-based algorithm using Hilbert series

Table 1 Timings $\mathbb{F}_{32003}$ in seconds

System	F5	F5E	GGV	GWV	New Alg.	CoCoA
Katsura 9	14.98	14.87	17.63	14.95	15.35	15.12
Katsura 10	153.35	152.39	192.20	152.32	165.32	142.21
Eco 8	2.24	0.38	0.49	0.69	0.35	0.44
Eco 9	77.13	8.19	13.51	8.23	7.21	13.26
F744	19.35	8.79	26.86	10.65	8.69	8.11
Cyclic 7	7.01	7.22	33.85	9.12	7.12	7.78
Cyclic 8	7,310.39	4,961.58	2,6242.12	5,961	6,267.58	4,410.39

Table 2 Number of critical pairs and zero reductions

System	F5		F5E		GGV		GWV		New Alg.	
Katsura 9	886	0	886	0	886	0	886	0	**886**	**0**
Katsura 10	1,781	0	1,781	0	1,781	0	1,781	0	**1,781**	**0**
Eco 8	830	322	565	57	2,012	57	602	57	**532**	**50**
Eco 9	2,087	929	1,278	120	5,794	120	1,778	120	**1,572**	**116**
F744	1,324	342	1,151	169	2,145	169	1,351	169	**1,091**	**162**
Cyclic 7	1,018	76	978	36	3,072	36	1,073	36	**978**	**30**
Cyclic 8	7,066	244	5,770	244	24,600	244	7,760	244	**7,592**	**239**

4 Experimental Results

In this section, we assess the performance of our algorithm against a standard database of examples to benchmark Gröbner basis algorithms.

We have implemented both GVW and new criteria in CoCoALib, and observed that both provide reasonable results. Also, the reader should keep in mind that this algorithm return more than just a Gröbner basis for the input ideal. All tests were run on Mac OS X with a 2.6 GHZ Intel Core i7 processor and 8GB DDR3 of memory.

The first benchmark compares the computation times of several examples over a field $\mathbb{F}_{32003}$. In order to ease the identification of our algorithm, we label it as *New Alg*. The computation times follow below:

We can draw two conclusions from this benchmark. First, the our algorithm is in general faster than the original algorithm—GWV, and in some cases it can be up to six times faster than GGV. Finally, we can see that there is not a clear winner but at comparable times our algorithm provides a Gröbner basis for the input ideal and a Gröbner basis for the leading terms of the syzygy module of the ideal.

Another way to compare the algorithm's performance is to use a computer independent measure, were we take into consideration the number of J-pairs processed and the number of J-pairs reducing to zero; therefore not producing any useful data. The data belonging to this measure can also be used to relate the amount of time required to compute the Gröbner basis (see Table 1) with the number of zero reductions predicted (Table 2).

5 Conclusions and Future Work

In this paper, we describe several strategies to improve the efficiency of the computation of a Gröbner basis. Despite any benefit that these approaches might have, they often have drawbacks, too. For example, one has often to consider restrictions on the input. Moreover, the efficiency of these methods is highly dependent of the behavior of the data during the computations, which cannot be known beforehand. Thus, these improvements do not provide us with a fail-proof optimization to compute Gröbner bases. To get a Gröbner basis in an efficient way, one has to implement and combine most of the presented ideas with well-designed heuristics.

The signature-based algorithms are not an exception. However, we can see that in most cases they will find more useless critical pairs than Buchberger's Criteria. Within the signature-based algorithms, performance is affected by restrictions on the reduction process and by the overhead that is generated due to how aggressive is the signature-based criteria chosen. Thus, again the question is not about getting a universal best algorithm, but even more about how to combine the signature-based world with already highly efficient improvements of the classic world without harming the performance or causing wrong results.

We developed a new strategy to the signature-based algorithms, which was built on top of previous literature concerning Hilbert series. There are two conclusions that can be drawn from the benchmarks. First, the our algorithm is in general faster than the original algorithm—GWV, and in some cases it can be up to six times faster than GGV. Finally, we can see that there is not a clear winner, but at comparable times our algorithm provides a Gröbner basis for the input ideal and a Gröbner basis for the leading terms of the syzygy module of the ideal.

Acknowledgments The author would like to thank to Prof. Lorenzo Robbiano for his thought provoking lectures and valuable comments. This work has been funded by the Marie-Curie Network SAGA (FP7, GA no. 214584).

References

1. Atiyah, M.F., MacDonald, I.G.: Introduction to Commutative Algebra. Addison-Wesley, London (1969)
2. Bigatti, A.M.: Computation of hilbert-poincaré series. J. Pure Appl. Algebra **119**(3), 237–253 (1997)
3. Buchberger, B.: An Algorithm for Finding the Bases Elements of the Residue Class Ring Modulo a Zero Dimensional Polynomial Ideal (German). Univ. of Innsbruck, Austria (1965)
4. Eder, C.: On the criteria of the f5 algorithm. In: arXiv:0804.2033v1 (2008)
5. Eder, C., Perry, J.: F5c: a variant of faugre's f5 algorithm with reduced gröbner bases. J Symbolic Comput. **45**(12), 1442–1458 (2010)
6. Eisenbud, D.: Commutative Algebra: With a View Toward Algebraic Geometry. Springer, Berlin (2008)
7. Faugre, J.C.: A new efficient algorithm for computing gröbner bases without reduction to zero (f5). In: Proceedings of ISSAC'02, pp. 75–82. ACM Press, New York (2002)

8. Gao, S., Guan, Y., Volny, F.I.: A new incremental algorithm for computing gröbner bases. In: Proceedings of the 2010 International Symposium on Symbolic and Algebraic Computation, ISSAC '10, pp. 13–19. ACM, New York, NY (2010). doi:10.1145/1837934.1837944. http://doi.acm.org/10.1145/1837934.1837944
9. Gao, S., Volny, F.I., Wang, M.: A new algorithm for computing gröbner bases. In: Cryptology ePrint Archive: Report (2010)
10. Greuel, G., Pfister, G.: A Singular Introduction to Commutative Algebra. Springer, Berlin (2007)
11. Hashemi, A., Ars, G.: Extended f5 criteria. J Symbolic Comput. **45**(12), 1330–1340 (2010)
12. Hoffmann, C.M.: Geometric and Solid Modeling: An Introduction. Morgan Kaufmann Publishers Inc., San Francisco, CA, USA (1989)
13. Hoffmann, C.M.: A dimensionality paradigm for surface interrogations. Comput. Aided Geom. Des. **7**(6), 517–532 (1990)
14. Kreuzer, M., Robbiano, L.: Computational Commutative Algebra. Springer, Berlin (2000)
15. Kreuzer, M., Robbiano, L.: Computational Commutative Algebra, 0th edn. Springer, Berlin (2005)
16. Kreuzer, M., Robbiano, L.: Computational Commutative Algebra, 2nd edn. Springer, Berlin (2009)
17. Stegers, T.: Faugres F5 algorithm revisited (2005)
18. Traverso, C.: Hilbert functions and the buchberger algorithm. J. Symb. Comput. **22**(4), 355–376 (1996). doi:10.1006/jsco.1996.0056. http://dx.doi.org/10.1006/jsco.1996.0056
19. Wilkinson, J.H.: The evaluation of the zeros of ill-conditioned polynomials. part i. Numerische Mathematik **1**, 150–166 (1959)

Numerical Algebraic Geometry and Differential Equations

Wenrui Hao, Bei Hu and Andrew J. Sommese

Abstract In this paper we review applications of numerical algebraic geometry to differential equations. The techniques we address are direct solution, bootstrapping by filtering, and continuation and bifurcation. We review differential equations systems with multiple solutions and bifurcations.

Keywords Numerical algebraic geometry · Polynomial system · Witness sets · Membership test · Numerical irreducible decomposition

Mathematics Subject Classification (2010) 65H10 · 68W30

1 Introduction

Most systems of nonlinear differential equations are not solvable in an explicit form, nor is the structure of the solutions; the number of solutions; or even the existence of a single solution with given boundary conditions. For these reasons, users of mathematics are forced to resort to numerical methods, though for nonlinear systems, traditional numerical methods usually do not suffice.

In recent years, remarkable progress has been made in the development and implementation of efficient algorithms to numerically solve and manipulate solutions of systems of polynomial equations. For some background on this field, called Numer-

W. Hao (✉) · B. Hu · A. J. Sommese
Department of Applied and Coumputational Mathematics and Statistics,
University of Notre Dame, Notre Dame, IN 46556, USA
e-mail: whao@nd.edu

B. Hu
e-mail: b1hu@nd.edu

A. J. Sommese
e-mail: sommese@nd.edu

R. De Amicis and G. Conti (eds.), *Future Vision and Trends on Shapes, Geometry and Algebra*, Springer Proceedings in Mathematics & Statistics 84,
DOI: 10.1007/978-1-4471-6461-6_3, © Springer-Verlag London 2014

ical Algebraic Geometry, see [30, 35, 38]. Many systems, when discretized, lead to systems of polynomials. In the last few years, methods of numerical algebraic geometry have begun to be used to investigate and solve systems of discretized nonlinear differential equations.

In this article, we will give an overview of the new approach and some of the systems that have been successfully investigated and solved by these new methods.The new approach has been based on combinations of three techniques:

1. direct solution of the polynomial systems that arise [18] and [28, Sect. 9.4];
2. bootstrap methods [1, 23] to solve large systems by building up to them from smaller systems; and
3. construction of new branches of solutions out of existing branches of solutions by computation of bifurcation points and continuation of solutions along the distinct branches coming out of the bifurcation points [19–22, 24–26].

2 Direct Solution

In this section we give some nontrivial examples of systems, where direct solution has been used. The combination of efficient new parallel algorithms; increasingly inexpensive clusters; and the polynomial solver Bertini [4] has led to the ability to solve discretizations with over forty polynomials in forty variables.

Given the exponential growth of the number of numerical solutions of the polynomial systems that arise through discretization as the number of nodes increase, the direct methods by themselves are hopeless. Nevertheless brute-force methods do allow significant information to be computed for a number of nontrivial systems, and combined with the bootstrap approaches explained in Sects. 3 and 4 allow us to compute many solutions for several nontrivial discretizations with thousands of variables.

2.1 A Lotka–Volterra Population Model with Diffusion

As a first example, consider the following Lotka–Volterra population model with diffusion [32, 37] on the first quadrant square $\mathscr{R} := [0, 1] \times [0, 1] \subset \mathbb{R}^2$:

$$- \Delta u = u(1 - v) \tag{1}$$

$$-\Delta v = -v(1 - u), \tag{2}$$

where Δ is the Laplacian and u and v are functions defined on $\mathscr{R}$, which are taken some non-zero boundary condition. For example,

$$u = \begin{cases} 0 & \text{if } x = 0 \\ 1 & \text{if } x = 1 \\ x & \text{if } y = 0, \\ \sin(x)/\sin(1) & \text{if } y = 1, \end{cases} \quad \text{and } v = \begin{cases} 1 & \text{if } x = 0, \\ 0 & \text{if } x = 1, \\ 1 - x & \text{if } y = 0, \\ \sin(1-x)/\sin(1) & \text{if } y = 1. \end{cases}$$

Discretizing using central differences with the gridpoints

$$(x_i, y_j) = \left(\frac{i}{5}, \frac{j}{n+1}\right) \quad \text{for } 0 \le j \le 5; 0 \le i \le n$$

for some positive integer n, we have the polynomial system with $8n$ equations

$$\frac{u_{i+1,j} - 2u_{i,j} + u_{i-1,j}}{25} + \frac{u_{i,j+1} - 2u_{i,j} + u_{i,j-1}}{(n+1)^2} + \frac{u_{i,j}(1 - v_{i,j})}{25(n+1)^2} = 0 \quad (3)$$

$$\frac{v_{i+1,j} - 2v_{i,j} + v_{i-1,j}}{25} + \frac{v_{i,j+1} - 2v_{i,j} + v_{i,j-1}}{(n+1)^2} + \frac{v_{i,j}(u_{i,j} - 1)}{25(n+1)^2} = 0 \quad (4)$$

with variables $\{u_{i,j}, v_{i,j} \mid 1 \le i \le n, 1 \le j \le 4\}$ and constants

$$u_{0,j} = v_{n+1,j} = 0, \; u_{n+1,j} = v_{0,j} = 1,$$

$$u_{i,0} = i/5, \; u_{i,5} = \frac{\sin(i/5)}{\sin(1)}, \; v_{i,0} = 1 - i/5, \; v_{i,5} = \frac{\sin(1 - i/5)}{\sin(1)}.$$

This system, which was investigated numerically by Hauenstein, Hu, and Sommese, is a good example of the growth of solutions of the discretization as the gridsize grows. In [28, Sect. 9.4] a related (and practically equivalent) polynomial system was used as one of the test examples for the polynomial-system solving method called regeneration. All equations are quadratic with total degree 2^{8n}, though the system has only 2^{4n} solutions. Bertini took 7.36 min when $n = 4$ and 3.81 h when $n = 5$ when run in parallel on 64 cores (8 nodes each with two 2.33 GHz quad-core Xeon 5410s). Though this grid is fine enough to get some useful information, the computing times make clear that (even with more cores), the brute force approach is limited. The main numerical polynomial system solvers Bertini [4], HOM4PS-2.0 [29], and PHC [29] were compared on the system of [28, Sect. 9.4]: only Bertini could deal with the larger systems.

Nevertheless direct computation can be very useful in some cases. The patterning model of Zhang et al. [39] leads to a system of four nonlinear ordinary differential equations of the form (5). We refer to [18], where this system of differential equations is written out in detail and explained (there are biologically relevant constants). In non-dimensionalized form the system was shown to be [18, Eq. 3.6]

$$\begin{cases} \dfrac{\partial A}{\partial T} = \dfrac{\partial^2 A}{\partial X^2} + H_1(A, B, C, S, T); \\ \dfrac{\partial B}{\partial T} = H_2(A, B); \\ \dfrac{\partial C}{\partial T} = \dfrac{\partial^2 C}{\partial X^2} + H_3(A, S, C); \\ \dfrac{\partial S}{\partial T} = \dfrac{\partial^2 S}{\partial X^2} + H_4(A, B, C, T), \end{cases} \tag{5}$$

where $0 \leq X \leq 1$ and the H_i are some specific (but involved) nonlinear functions. Using an interval with eleven points, the associated polynomial system was solved using Bertini. This took about 11 h using 200 cores (25 nodes each with two 2.33 GHz quad-core Xeon 5410s). The associated polynomial system had 384,064 complex solutions of which 17,974 were real. Of these, only seven satisfied the physically necessary condition that they are positive. Three of the seven were shown to be stable. Many of the seven solutions were new and would have been very difficult to compute by standard methods such as time-marching.

3 Bootstrapping by Filtering

Given the systems in Sect. 2, we see that for polynomial system methods to be of broad use, we need to be able to cut the exponential growth of solutions as the number of grid points grows. One way to do this [1–3] is *filtering*. To explain this approach consider the example following ordinary differential equation on [0, 1], which is know to have infinitely many solutions [1, Sect. 3.3]:

$$y'' = -y^3. \quad y(0) = y(1) = 0. \tag{6}$$

Discretized we have the polynomial systems F_N indexed by integers $N \geq 1$

$$F_N(y) := \begin{bmatrix} \dfrac{y_2 - 2y_1}{h^2} + y_1^3 \\ \dfrac{y_3 - 2y_2 + y_1}{h^2} + y_2^3 \\ \vdots \\ \dfrac{y_N - 2y_{N-1} + y_{N-2}}{h^2} + y_{N-1}^3 \\ \dfrac{-2y_N + y_{N-1}}{h^2} + y_N^3 \end{bmatrix} = 0 \tag{7}$$

for $N = 1, 2, 3, \ldots$, where $h = 1/(N+1)$, $y_0 = y_{N+1} = 0$, and for $1 \le i \le N+1$, y_i is the approximate value of a given possible solution at $x_i = i/(N+1)$. Here we have presented the system as if $N \ge 4$ so that the pattern is clear.

Of course, we cannot numerically compute all solutions since there are an infinite number, but we would like to compute a good selection of solutions for increasing N. Only a small fraction (see [1]) of the solutions for any N are real.

The idea of filtering is to start with some set $\mathscr{U}_N$ of isolated solutions of $F_N(y) = 0$. To get to solutions of $F_{N+1}(y) = 0$

1. use some filtering condition (for example, having small imaginary parts) to discard solutions, leaving a set $\mathscr{S}_N$ of solutions of $F_N(y) = 0$;
2. construct a polynomial system in $y_1, \ldots, y_{N+1}$ of the form

$$\Phi(y_1, \ldots, y_N, y_{N+1}) = \begin{bmatrix} F_N(y_1, \ldots, y_N) \\ g(y_1, \ldots, y_N, y_{N+1}) \end{bmatrix} = 0;$$

3. compute the solutions $\mathscr{I}_{N+1}$ of $\Phi(y_1, \ldots, y_{N+1}) = 0$ with $(y_1, \ldots, y_N) = y^*$ for some $y^* \in \mathscr{S}_N$;
4. construct a homotopy $H(y_1, \ldots, y_{N+1}, t) = 0$ with $t \in [0, 1]$; $H(y, 1) = \Phi(y)$; and $H(y, 0) = F_{N+1}(y)$;
5. use $H(y, t)$ to continue the solutions $\mathscr{I}_{N+1}$ to solutions $\mathscr{U}_{N+1}$ of $F_{N+1} = 0$.

This process will typically start with all solutions of $F_{N_0} = 0$ for a small integer N_0. Clearly there are a lot of choices. Moreover we might add more nodes at each step.

4 Bootstrapping by Domain Decomposition

Though filtering works well with ordinary differential equations, it, by itself, has not worked well with systems of nonlinear partial differential equations. In [23], we introduce a new bootstrapping method to use domain decomposition to guide us in building up to a polynomial system (arising by discretization) from simpler polynomial systems (arising from coarser discretizations and discretizations over smaller regions). This new method combines well with filtering to yield many unknown solutions of interesting systems. In [23, Sect. 5], this method is used to yield many highly nontrivial solutions of complicated tumor growth models of the sort discussed in Sect. 5. Here we just give a simple illustrative ordinary differential equation example.

Consider the system

$$u_{xx} = f(u) \text{ on } (0, 1); \quad u(0) = u_0, \ u(1) = u_1 \tag{8}$$

for some polynomial $f(u)$: in the article [23], a number of different $f(u)$ are used in examples.

Our goal is to first solve Eq. 8 on the interval [0, 1] with NM grid points where N and M are two small integers. To do this

1. we first solve Eq. 8 for the gridpoints at $x_i = iH$, where $i = 0, \ldots, N$ and $H = 1/N$ and use some filter to discard unreasonable solutions;
2. next we solve
$$u_{xx} = f(u)$$
with M gridpoints on each of the intervals $[x_i, x_{i+1}]$ with boundary conditions given by using the solutions in step (1); and
3. we use a homotopy to continue the solutions of step (2) into solutions of a discretization of Eq. 8 with NM gridpoints.

We first solve

$$C_N(u_1, \ldots, u_{N-1}) = \begin{bmatrix} u_2 - 2u_1 + u_0 & - H^2 f(u_1) \\ & \vdots \\ u_N - 2u_{N-1} + u_{N-2} & - H^2 f(u_{N-1}) \end{bmatrix} = 0. \quad (9)$$

Let $x_{i,j} = x_i + jh$ where $h = 1/M$ and $j = 0, \ldots, M$ Note that $x_{i,0} = x_i$ and $x_{i,M} = x_{i+1}$. We wish to compute an approximation $u_{i,j}$ to a solution $u(x_{i,j})$ of Eq. 8 using the discretization with $NM - 1$ variables and equations

$$\mathscr{F}_{NM}(U) = \begin{bmatrix} u_{0,2} - 2u_{0,1} + u_{0,0} - (hH)^2 f(u_{0,1}) \\ \vdots \\ u_{N-1,M} - 2u_{N-1,M-1} + u_{N-1,M-2} - (hH)^2 f(u_{N-1,M-1}) \end{bmatrix} = 0 \quad (10)$$

where we let U denote $(u_{0,1}, \ldots, u_{0,M-1}, u_{1,0}, \ldots, u_{1,0}, \ldots, u_{N-1,0}, \ldots, u_{N-1,M-1})$ and make the convention that the variables $u_{i,N}$ and $u_{i+1,0}$ are the same for $i = 0, \ldots, N-1$, $u_{N-1,M} = u(1)$, and $u_{0,0} = u(0)$. For each solution $(u_0, \ldots, u_N)$ of Eq. 9, compute approximate solutions to the systems

$$u_{xx} = f(u) \text{ on } [x_i, x_{i+1}]; \;\; u(x_i) = u_i, \;\; u(x_{i+1}) = u_{i+1}$$

using the systems

$$SS_{i,M}(u_{i,0}, u_{i,1}, \ldots, u_{i,M-1}, u_{i,M}) = \begin{bmatrix} u_{i,2} - 2u_{i,1} + u_{i,0} & - h^2 f(u_{i,1}) \\ & \vdots \\ u_{i,M} - 2u_{i,M-1} + u_{i,M-2} & - h^2 f(u_{M-1}) \end{bmatrix} = 0 \quad (11)$$

for $i = 0, \ldots, N-1$ with $u_{0,0} = u_0$ and $u_{N-1,M} = u_1$.

We thus have a set of solutions of the composite system

$$P_N(U) = \begin{bmatrix} C_N(u_{0,M}, u_{1,M}, \ldots, u_{N-1,M}) \\ SS_{0,M}(u_{0,0}, u_{0,1}, \ldots, u_{0,M}) \\ \vdots \\ SS_{N-1,M}(u_{N-1,0}, \ldots, u_{N-1,M}) \end{bmatrix} = 0. \tag{12}$$

Next we track these solutions as t goes from 1 to 0 using a homotopy such as

$$H(U,t) = (1-t)\mathscr{F}_{NM}(u_{0,1}, \ldots, u_{N-1,M-1}) + t \begin{bmatrix} C_N(u_1, \ldots, u_{N-1}) \\ SS_{0,M}(u_{0,0}, \ldots, u_{0,M}) \\ \vdots \\ SS_{N-1,M}(u_{N-1,0}, \ldots, u_{N-1,M}) \end{bmatrix}.$$

Finally filtering the solutions of $H(U, 0)$.

5 Continuation and Bifurcation

In this section we only discuss the free boundary problems arising in tumor growth [19–23] that we have investigated using the new methods. These methods apply to other biological models [24, 25].

The free boundary problems are of the type

$$V_n(x,t) = F[\mathscr{O}(t), \mathbf{u}(x,t), \lambda], \quad x \in \Gamma(t), \quad t > 0, \tag{13}$$

where $\mathbf{u}(x,t)$ is typically a solution of a system of partial differential equations in the domain $x \in \mathscr{O}(t), t > 0$ which is changing over time, $V_n(x,t)$ $(x \in \Gamma(t))$ is the normal velocity of the boundary of the domain $\Gamma(t) = \partial\mathscr{O}(t)$, and F is a functional of the domain $\mathscr{O}(t)$ and the function $\mathbf{u}$. Typically, F is a nonlinear functional given by the derivatives of $\mathbf{u}$ on the boundary $\Gamma(t)$. Here the parameter λ represents various physical quantities that may change the behavior of the system.

Let us see how such problems arise in the growth of a solid tumor. Consider a (very much simplified) biological model of a tumor. In this simplified model, oxygen and glucose are considered "nutrient", with its density c satisfying $\delta c_t - \Delta c = -c$ for $x \in \mathscr{O}(t)$. Since the rate of diffusion of a nutrient is much faster than the rate of cell proliferation, it is also reasonable to take δ to be zero (Quasi-steady state approximation). The tumor grows with proliferation rate $= \mu(c - \tilde{c})$, where $\tilde{c}$ is a threshold concentration and μ is a parameter expressing the "intensity" of the proliferation (if $c > \tilde{c}$) or shrinkage by necrosis (if $c < \tilde{c}$) within the tumor. By the conservation of mass, proliferation rate $= \nabla \cdot \mathbf{v}$, where $\mathbf{v}$ is the velocity field within the tumor. If the tissue is assumed to be of porous medium type where Darcy's law (i.e., $\mathbf{v} = -\nabla\sigma$, where σ is the pressure) can be used (here the extracellular matrix is considered *"porous media"* in which the cell moves), then $-\Delta\sigma = \mu(c - \tilde{c})$, and the system is reduced to finding two unknown functions c and σ, together with the free boundary $\Gamma(t)$:

$$\begin{array}{ll} \delta c_t - \Delta c + c = 0, & x \in \mathcal{O}(t), \\ c = \bar{c}, & x \in \Gamma(t), \\ -\Delta\sigma = \mu(c - \tilde{c}), & x \in \mathcal{O}(t), \\ \sigma = \gamma\kappa, & x \in \Gamma(t), \\ V_n(t) = -\frac{\partial\sigma}{\partial n}, & x \in \Gamma(t), \end{array} \tag{14}$$

where $\bar{c}$ represents the concentration of the nutrient surrounding the tumor.

Models such as that above are mathematically extremely difficult. For example, if we set μ to be zero, then this problem reduces to the classical Hele-Shaw problem with surface tension (with many works devoted to it in the literature—searching Hele-Shaw on the title alone on MathSciNet returns 390 entries). Naturally, the tumor aggressiveness constant μ cannot be zero; thus this problem is more complex than the Hele-Shaw problem and the classical existence of a solution is in general not expected to be global in time. Fingering phenomenon is well observed for Hele-Shaw models. For our model, one of the most important questions is whether the tumor will spread out of control, or remains bounded. Since the study of the general classical existence (globally in time) becomes unrealistic (it is known that for Hele-Shaw with surface tension, global classical existence is not expected globally in time), it is natural to study the radially symmetric case as the first step. Although the tumor in vivo is unlikely to be radially symmetric, the tumor in vitro grown in a laboratory is likely to be of spherical shape. Thus the model does have implications in the application.

The study of this model started in the 1990s. In particular, if the tumor aggressiveness constant μ is large, or if the cell-to-cell adhesiveness constant γ is small, then the tumor is likely to spread. The radially symmetric case was studied in [15], where they established rigorously the global classical existence of a solution under natural biological assumptions; they also established the stability results: (1) there is a stationary solution under the natural biological data, (2) for small μ, the stationary solution is asymptotically stable with respect to radially symmetric small perturbations. Note that this stability result is expected biologically since, as indicated above, μ describes the aggressiveness of the tumor.

It is natural to ask what happens to a non-radially symmetric solution. As mentioned above, the general study of non-radially symmetric solutions is difficult. Thus, as a first step, we like to start with a question that is simpler and yet important enough for the application: does a non-radially symmetric solution exist? This question is answered positively in [16], where they linearize the problem around a radially symmetric solution and then they formed an analytic series expansion near the free boundary. For the 2-space dimensional case, through a careful study using sharp PDE estimates, they proved that the series is convergent, and thus completing the proof of the existence of non-radially symmetric solution. The proof is limited to 2-space dimensional case since the PDE estimates are very lengthy and complex—they are of the type of the Cauchy–Kowalevski theorem. In [8], Hanzawa transformation is used to simplify the proof and extend the result to 3-space dimensional case.

In a recent series of papers [12–14] Friedman and Hu have developed bifurcation theories and stability theorems, combining Crandall–Rabinowitz theorem [5] with new estimates on the PDEs in addition to the fundamental PDE estimates [6, 17, 31]. The methods involve *very sharp estimates* on the solutions of PDEs, using dimension reduction by Laplace transform, and explicit expansion into series of spherical harmonics. They have considered the stationary solutions and proved, that given any $R > 0$, one can construct branches of stationary solutions with any number of fingers, that is, solutions with free boundary

$$r = R + \varepsilon Y_{n,0}(\theta) + O(\varepsilon^2), \qquad n \geq 2, \qquad \mu = \mu_n(R) + \mu_{n,0}(R)\varepsilon + O(\varepsilon^2), \quad (15)$$

(here we assume $\gamma = 1$ for simplicity), for any small ε. In this model μ_j are monotone: $\mu_2(R) < \mu_3(R) < \mu_4(R) < \cdots$, for any $R > 0$. It is clear that the stability of the stationary solution is lost at the first bifurcation point $\mu = \mu_2(R)$. Friedman and Hu [12] actually established the stability result for the stationary solution for all $\mu < \mu_*(R)$ (here R is the radius of the stationary solution) and non-stability result for $\mu > \mu_*(R)$. Furthermore, they rigorously proved that $\mu_*(R) < \mu_2(R)$ for $R < \bar{R}$ and $\mu_*(R) = \mu_2(R)$ for $R \geq \bar{R}$, for some critical $\bar{R}$ which can be computed numerically in terms of the tumor aggressiveness constant μ and the cell-to-cell adhesiveness constant γ. (It is also not surprising that the bifurcation result when γ is not normalized to 1 is given in terms of the ratio μ/γ.)

Of course, the bifurcation rigorously established in (15) describes only the behavior of stationary solution near the bifurcation point (i.e., $|\varepsilon|$ small). In reality, it is interesting to find out what happens for the bifurcation branch away from the bifurcation point, and this is where the numerical computation is needed: to find, by a homotopy method (a frequently used method in the classical study of PDEs), the steady state branches of solutions as the data moves away from the bifurcation point. The next task will be to find out the stability of these solutions, as this will give a strong hint as to whether the tumor will spread out of control.

While Darcy's law may be a good approximation for some solid tumors, for several models (ductal carcinoma in breast, brain tumor) the Stokes equation is more appropriate [9–11]. In this case the stress tensor is given by $\sigma_{ij} = -\sigma\delta_{ij} + 2\nu\left(e_{ij} - \frac{1}{3}\bar{\Delta}\,\delta_{ij}\right)$ where $\sigma = -\frac{1}{3}\sigma_{kk}$, $e_{ij} = \frac{1}{2}\left(\frac{\partial v_i}{\partial x_j} + \frac{\partial v_j}{\partial x_i}\right)$ is the strain tensor, $\bar{\Delta} = e_{kk} = \mathrm{div}\mathbf{v}$ is the dilation, and ν is the viscosity coefficient. If there are no body forces then $\frac{\partial \sigma_{ij}}{\partial x_j} = 0$, which can be written as the Stokes equation

$$-\nu\Delta\mathbf{v} + \nabla\sigma - \frac{1}{3}\nu\nabla\mathrm{div}\mathbf{v} = 0, \quad x \in \mathscr{O}(t), \quad t > 0. \quad (16)$$

Assuming that the strain tensor is continuous up to the boundary of the domain, we then obtain a boundary condition:

$$T\mathbf{n} = -\gamma\kappa\mathbf{n}, \quad x \in \Gamma(t), \quad t > 0,$$

where T is the stress tensor: $T = \nu(\nabla\mathbf{v} + (\nabla\mathbf{v})^T) - (\sigma + \frac{2}{3}\nu \operatorname{div}\mathbf{v})I$ with components

$$T_{ij} = \nu\Big(\frac{\partial v_i}{\partial x_j} + \frac{\partial v_j}{\partial x_i}\Big) - \delta_{ij}\Big(p + \frac{2\nu}{3}\operatorname{div}\mathbf{v}\Big),$$

$\mathbf{n}$ is the outward normal, and κ is the mean curvature.

The free boundary condition is given by

$$V_n(t) = \mathbf{v} \cdot \mathbf{n}, \qquad x \in \Gamma(t). \tag{17}$$

Replacing, in the tumor model described above, Darcy's law by Stokes equation (16) while keeping the rest of the equations, we obtain a free boundary problem for a coupled system of Stokes equation and a diffusion equation.

The theoretical results established in [7, 12, 13] describe the behavior near the bifurcation point (i.e., $|\varepsilon|$ small). Again, it is interesting to find out what happens for the bifurcation branch when the data moves away from the bifurcation point, and this is where the numerical computation comes in: to find, by the method of continuation, the steady state solutions as the data moves away from the bifurcation point, and to find out the stability of these solutions.

The theoretical study of the above tumor problems depends on the explicit formula of the radially symmetric solution and sharp PDE estimates. Explicit solutions will not always be available.

For example, let the tumor region be $\mathscr{O}(t)$ and assume that there are several types of cells within the tumor: proliferating cells with density $p(x,t)$, quiescent cells with density $q(x,t)$. The nutrient density with the tumor is still denoted by $c(x,t)$. Proliferating cells change into quiescent cells at the rate $K_R(c)$, and they become necrotic at a rate $K_N(c)$. We use $K_B(c)$ to represents the balance between birth and death. Then, similar to Problem 1, the equations for conservation of mass are given by [33, 34]

$$\frac{\partial p}{\partial t} + \operatorname{div}(p\mathbf{v}) = [K_B(c) - K_Q(c)]p + K_R(c)q, \qquad x \in \mathscr{O}(t), \tag{18}$$

$$\frac{\partial q}{\partial t} + \operatorname{div}(q\mathbf{v}) = K_Q(c)p - [K_R(c) + K_N(c)]q, \qquad x \in \mathscr{O}(t), \tag{19}$$

where $\mathbf{v}$ is velocity of the cells within the tumor. This velocity is a result of the proliferation of cells and removal of necrotic cells. The nutrient $c(x,t)$ diffuses within the tumor and is therefore modeled by the diffusion equation, as before.

Consider the 2-space dimensional tumor. To handle the free boundary, we developed a moving grid that evolved with the boundary. Figure 1 shows the grid for a radially symmetric and a non-radially symmetric solution. Using a third order discretization scheme with $R = 2.5$, we computed the radially symmetric solution for a small random value of μ and then used parameter continuation implemented in Bertini to track this solution as μ varied.

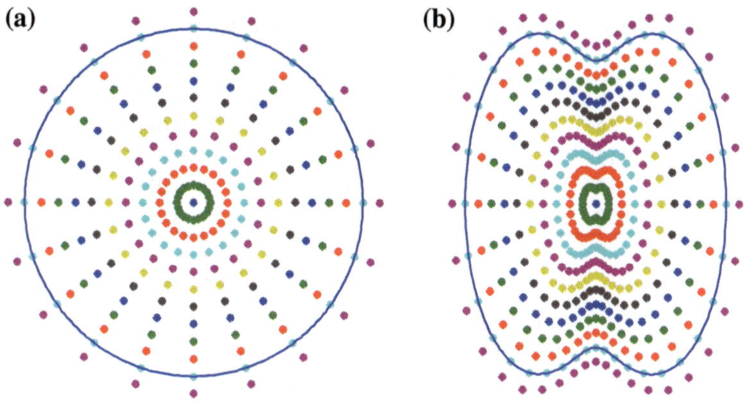

Fig. 1 Pictures of grids. **a** Radially symmetric. **b** Non-radially symmetric

Fig. 2 Condition number along radially symmetric solution branch as a function of μ

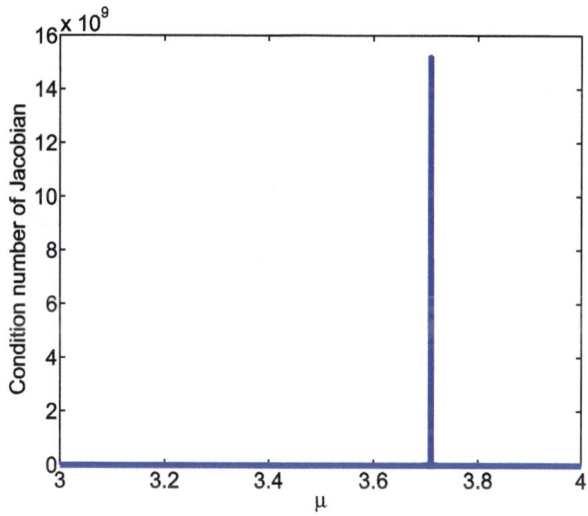

Since the Jacobian matrix of the discretized polynomial system is rank deficient when evaluated at the radially symmetric solution corresponding to a (discretized) bifurcation value of μ, we monitored the condition number of the Jacobian matrix as we tracked along the branch of radially symmetric solutions. This process produced a clear indication of a potential (discretized) bifurcation value near $\mu = 3.7$ as shown in Fig. 2. Further calculations, described below, indeed show that this value corresponds to an approximation of $\mu_2(R)$.

To produce a better approximation of the bifurcation value near $\mu = 3.7$, we setup Bertini to take increasingly smaller steps along the radially symmetric solution branch near $\mu = 3.7$. The numerically computed value matches the theoretical value very well, where the two-dimensional theoretical bifurcation value was derived in the same way as the three-dimensional case. The three-dimensional case was rigorous

Table 1 Comparing (discretized) bifurcation value of μ_2 on a sequence of grids

N_θ	N_R	μ_2	Abs. error
40	10	3.725819	0.023132
48	12	3.720450	0.017763
52	13	3.718400	0.015713
60	15	3.715204	0.012517
80	20	3.710412	0.007725

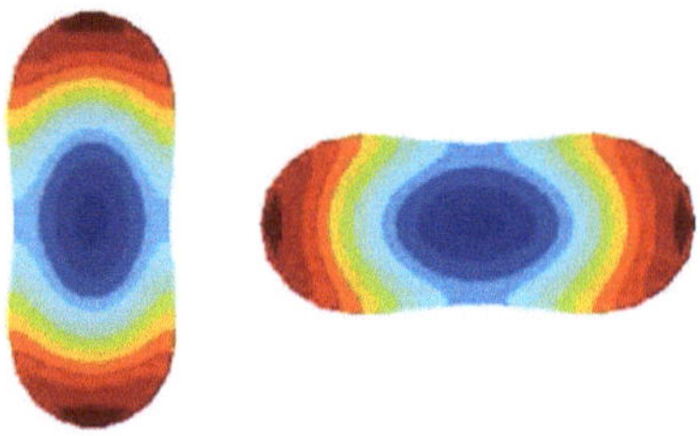

Fig. 3 Non-radially symmetric solutions

derived and presented along with a three-dimensional theoretical bifurcation diagram in [12–14]. Table 1 compares these bifurcation values on a sequence of grids with the theoretical value $\mu_2(R) = 3.702687$. In this table, N_θ and N_R are the number of grid points in the angular and radial directions, respectively. It is clear that the numerical value converges to the theoretical value in our experiment.

For these computations, the cost of numerically computing the radially symmetric solution is trivial compared with the cost of computing the condition number of the Jacobian matrix for the discretized system. All computations performed for this test are naturally parallelizable. For example, when using 200 processing cores, these computations with $N_\theta = 40$ and $N_R = 10$ took 3 min.

We verified the existence of a bifurcation branch by computing the tangent cone at these values. This produced two tangent directions which correspond to the radially symmetric branch and a symmetry-breaking branch. The numerically computed tangent direction for the symmetry-breaking branch compares favorably with its theoretical value, which was derived in the same way as the three-dimensional case discussed above. After computing the tangent direction of the symmetry-breaking branch, we forced Bertini to track along this branch. In Fig. 3, we present two solutions computed by tracking in opposite directions on this non-radially symmetric branch. Since the initial steps along this symmetry-breaking branch are poorly conditioned, using high precision to perform this computation is crucial.

To further verify our numerical computations, we used our most coarse discretization ($N_\theta = 40$ and $N_R = 10$) to approximate the value of μ_4. The discretized value computed was 18.624 which compares favorably with the theoretical value $\mu_4(R) = 18.649$.

Though the solutions, for the two and three-dimensional model problems we investigated, became nonconvex, the domains remain star-shaped with respect to the origin and they never reached the point where we had to switch to a more sophisticated grid. We expect that this will be the case for more complicated systems and, as needed, will utilize adaptive grid generation.

Building from the success in the 2-space dimensional case, we performed calculations in the 3-space dimensional case. Using spherical polar coordinates and a discretization with $N_R = 20$, $N_\theta = 20$, and $N_\phi = 40$ grid points in the radial, inclination, and azimuth directions, respectively, we approximated $\mu_2(R)$ where $R = 2.5$ as above. The discretized value computed was 4.0398 which compares favorably with the theoretical value of $\mu_2(R) = 4.0422$.

As noted above, the theoretical description of the non-radially symmetric solution branches are known only locally near the bifurcation point, that is, when $|\varepsilon|$ is small. A numerical homotopy allows one to compute data about the non-radially symmetric solution branch far away from the bifurcation point.

Another problem dealt with was deciding whether a solution of the polynomial system is a solution of the differential equation. This was accomplished by sharpening the solution to a higher precision; interpolating to a finer grid; refining the interpolated solution to a solution within the desired tolerance for the polynomial system associated to the finer grid; and finally comparing the solutions on the different grids. Deciding whether a point is a solution of a polynomial system is a basic and nontrivial part of this procedure, which has been dealt with in Bertini by using multiple levels of precision.

Using adaptive-precision continuation methods, it is straightforward to check nonlinear stability directly. For example, given a solution of the time-independent three-dimensional system, we took a random perturbation and checked the development of the system with respect to time to numerically verify the stability results of [12–14].

References

1. Allgower, E.L., Bates, D.J., Sommese, A.J., Wampler, C.W.: Solution of polynomial systems derived from differential equations. Computing **76**, 1–10 (2006)
2. Allgower, E.L., Cruceanu, S.G., Tavener, S.: Application of numerical continuation to compute all solutions of semilinear elliptic equations. Adv. Geom. **9**, 371–400 (2009)
3. Allgower, E.L., Cruceanu, S.G., Tavener, S.: Turning points and bifurcations for homotopies of analytic maps. Contemp. Math. **496**, 1–10 (2009)
4. Bates, D.J., Hauenstein, J.D., Sommese, A.J., Wampler, C.W.: Bertini: software for numerical algebraic geometry. Available at www.nd.edu/~sommese/bertini
5. Crandall, M.G., Rabinowitz, L.H.: Bifurcation from simple eigenvalues. J. Funct. Anal. **8**, 321–340 (1971)
6. Friedman, A.: Partial Differential Equations of Parabolic Type. Princeton-Hall, Englewood Cliffs (1964)
7. Friedman, A.: A free boundary problem for a coupled system of elliptic, hyperbolic, and Stokes equations modeling tumor growth. Interfaces Free Bound **8**, 247–261 (2006)

8. Fontelos, M., Friedman, A.: Symmetry-breaking bifurcations of free boundary problems in three dimensions. Asymptotic Anal. **35**, 187–206 (2003)
9. Franks, S.J.H., Byrne, H.M., King, J.P., Underwood, J.C.E., Lewis, C.E.: Modeling the early growth of ductal carcinoma in situ of the breast. J. Math. Biol. **47**, 424–452 (2003)
10. Franks, S.J.H., Byrne, H.M., King, J.P., Underwood, J.C.E., Lewis, C.E.: Modeling the growth of ductal carcinoma in situ. Math. Med. Biol. **20**, 277–308 (2003)
11. Franks, S.J.H., Byrne, H.M., Underwood, J.C.E., Lewis, C.E.: Biological inferences from a mathematical model of comedo ductal carcinoma in situ of the breast. J. Theor. Biol. **232**, 523–543 (2005)
12. Friedman, A., Hu, B.: Bifurcation from stability to instability for a free boundary problem arising in a tumor model. Arch. Rat. Mech. Anal. **180**, 293–330 (2006)
13. Friedman, A., Hu, B.: Asymptotic stability for a free boundary problem arising in a tumor model. J. Differ. Eqn. **227**, 598–639 (2006)
14. Friedman, A., Hu, B.: Stability and instability of Liapounov–Schmidt and Hopf bifurcation for a free boundary problem arising in a tumor model. Trans. Amer. Math. Soc. **360**, 5291–5342 (2008)
15. Friedman, A., Reitich, F.: Analysis of a mathematical model for growth of tumor. J. Math. Biol. **38**, 262–284 (1999)
16. Friedman, A., Reitich, F.: Symmetry-breaking bifurcation of analytic solutions to free boundary problems: an application to a model of tumor growth. Trans. Amer. Math. Soc. **353**, 1587–1634 (2000)
17. Gilbarg, D., Trudinger, N.S.: Elliptic Partial Differential Equations of Second Order. Springer, New York (1983)
18. Hao, W., Hauenstein, J.D., Hu, B., Liu, Y., Sommese, A.J., Zhang, Y.-T.: Multiple stable steady states of a reaction-diffusion model on zebrafish dorsal-ventral patterning. Discrete Continuous Dyn. Syst. Ser. S **4**, 1413–1428 (2011)
19. Hao, W., Hauenstein, J.D., Hu, B., Sommese, A.J.: A three-dimensional steady-state tumor system. Appl. Math. Comput. **218**, 2661–2669 (2011)
20. Hao, W., Hauenstein, J.D., Hu, B., Liu, Y., Sommese, A.J., Zhang, Y.-T.: Bifurcation for a free boundary problem modeling the growth of a tumor with a necrotic core. Nonlinear Anal. Ser. B Real World Appl. **13**, 694–709 (2012)
21. Hao, W., Hauenstein, J.D., Hu, B., Liu, Y., Sommese, A.J., Zhang, Y.T.: Continuation along bifurcation branches for a tumor model with a necrotic core. J. Sci. Comput. (to appear)
22. Hao, W., Hauenstein, J.D., Hu, B., McCoy, T., Sommese, A.J.: Computing steady-state solutions for a free boundary problem modeling tumor growth by Stokes equation. J. Comput. Appl. Math. **237**, 326–334 (2013). doi:10.1016/j.cam.2012.06.001
23. Hao, W., Hauenstein, J.D., Hu, B., Sommese. A.J.: A domain decomposition algorithm for computing multiple steady states of differential equations. Available at www.nd.edu/~sommese/preprints
24. Hao, W., Hu, B., Sommese, A.J.: Cell cycle control and bifurcation for a free boundary problem modeling tissue growth
25. Hao, W., Lin, G., Xu, Z., Rosen, E., Sommese, A.J., Alber, M.: Effect of fitted reaction rates on predicting thrombin production using blood coagulation model
26. Hao, W., Hauenstein, J.D., Shu, C.-W., Sommese, A.J., Xu, Z., Zhang, Y.-T.: A homotopy method based on WENO schemes for solving steady state problems of hyperbolic conservation laws
27. Hauenstein, J.D., Sommese, A.J.: Witness sets of projections. Appl. Math. Comp. **217**(7), 3349–3354 (2010)
28. Hauenstein, J.D., Sommese, A.J., Wampler, C.W.: Regeneration homotopies for solving systems of polynomials. Math. Comp. **80**, 345–377 (2011)
29. Lee, T.-L., Li, T.Y., Tsai, C.-H.: HOM4PS-2.0, solving polynomial systems by the polyhedral homotopy method. Software available at www.math.msu.edu/~li
30. Li, T.-Y.: Numerical solution of polynomial systems by homotopy continuation methods. In: Handbook of Numerical Analysis, vol. XI, pp. 209–304. North-Holland Press, Amsterdam (2003)

31. Lieberman, G.M.: Second Order Parabolic Differential Equations. World Scientific, Singapore (1996)
32. Lotka, A.J.: Undamped oscillations derived from the laws of mass action. J. Amer. Chem. Soc. **42**, 1595–1599 (1920)
33. Pettet, G.J., Please, C.P., Tindall, M.J., McElwain, D.L.S.: The migration of cells in multicell tumor spheroids. Bull. Math. Biol. **63**, 231–257 (2001)
34. Ribba, B., Colin, T., Schnell, S.: A multiscale mathematical model of cancer, and its use in analyzing irradiation therapies. Theor. Biol. Med. Model. **3**, 1–19 (2006)
35. Sommese, A.J., Wampler, C.W.: The Numerical Solution of Systems of Polynomials Arising in Engineering and Science. World Scientific, Singapore (2005)
36. Verschelde, J.: Algorithm 795: PHCpack: a general-purpose solver for polynomial systems by homotopy continuation. ACM Trans. Math. Software **25**(2)), 51–276 (1999) (Software available at www.math.uic.edu/~jan)
37. Volterra, V.: Variazionie fluttuazioni del numero dindividui in specie animali convivent. Mem. Acad. Lincei. **2**, 31–113 (1926)
38. Wampler, C.W., Sommese, A.J.: Numerical algebraic geometry and algebraic kinematics. Acta Numerica **20**, 469–567 (2011)
39. Zhang, Y.-T., Lander, A., Nie, Q.: Computational analysis of BMP gradients in dorsal-ventral patterning of the zebrafish embryo. J. Theor. Biol. **248**, 579–589 (2007)

Sparse Discriminants and Applications

Ioannis Z. Emiris and Anna Karasoulou

Abstract Polynomial algebra offers a standard, powerful, and robust approach to handle several important problems in geometric modelling and other areas. A key tool is the discriminant of a univariate polynomial, or of a well-constrained system of polynomial equations, which expresses the existence of multiple (or degenerate) roots. We describe discriminants in a general context, and relate them to an equally useful object, namely the resultant of an overconstrained polynomial system. We discuss several relevant applications in geometric modelling so as to motivate the use of such algebraic tools in further geometric problems. We focus on exploiting the sparseness of polynomials via the theory of Newton polytopes and sparse elimination.

Keywords Discriminant · Geometric application · Resultant · Sparse polynomial

Mathematics Subject Classification (2010) 13P15

1 Introduction

Polynomial algebra offers a standard approach to handle several problems in geometric modeling and other fields, which provides both powerful and robust methods. Polynomials arise in a variety of scientific and engineering applications, and can be manipulated either algebraically or numerically. Here we mostly focus on tools relevant for algebraic computation, but also useful in numerical calculations. In particular, the study and solution of systems of polynomial equations has been a major topic. Discriminants is a key tool when examining well-constrained systems,

I. Z. Emiris (✉) · A. Karasoulou
Department of Informatics and Telecommunications, University of Athens, Athens, Greece
e-mail: emiris@di.uoa.gr

A. Karasoulou
e-mail: akarasou@di.uoa.gr

R. De Amicis and G. Conti (eds.), *Future Vision and Trends on Shapes, Geometry and Algebra*, Springer Proceedings in Mathematics & Statistics 84,
DOI: 10.1007/978-1-4471-6461-6_4, © Springer-Verlag London 2014

including the case of one univariate polynomial. Their theoretical study is a thriving and fruitful domain today, but they are also very useful in a variety of applications. Through the related software development, these algebraic tools can be applied in various practical questions.

The best studied discriminant is that of one polynomial in one variable, probably known since high school, where one studies the discriminant of a quadratic polynomial $f(x) = ax^2 + bx + c, a \neq 0$. Polynomial f has a double (real) root if and only if its discriminant

$$\Delta = b^2 - 4ac$$

is equal to zero. Equivalently, this can be defined as the condition for $f(x)$ and its derivative $f'(x)$ to have a common root:

$$\exists x\colon f(x) = ax^2 + bx + c = f'(x) = 2ax + b = 0 \Leftrightarrow \Delta = 0. \tag{1}$$

One can similarly consider the discriminant of a univariate polynomial of any degree. If we wish to calculate the discriminant of a polynomial f of degree five in one variable, we consider the condition that both polynomial and its derivative vanish:

$$f(x) = ax^5 + bx^4 + cx^3 + dx^2 + ex + g = 0,$$
$$f'(x) = 5ax^4 + 4bx^3 + 3cx^2 + 2dx + e = 0.$$

In this case, elimination theory reduces the computation of discriminant Δ to the computation of a 9×9 Sylvester determinant, expressing the resultant of f, f'. If we develop this determinant, we encounter an instance of the fact that the number of its monomials increases exponentially with the input degree:

$$\begin{aligned}
\Delta = {} & -2050a^2g^2bedc + 356abed^2c^2g - 80b^3ed^2cg + 18dc^3b^2ge \\
& - 746agdcb^2e^2 + 144ab^2e^4c - 6ab^2e^3d^2 - 192a^2be^4d - 4d^2ac^3e^2 \\
& + 144d^2a^2ce^3 - 4d^3b^3e^2 - 4c^3e^3b^2 - 80abe^3dc^2 + 18b^3e^3dc \\
& + 18d^3acbe^2 + d^2c^2b^2e^2 - 27b^4e^4 - 128a^2e^4c^2 + 16ac^4e^3 - 27a^2d^4e^2 \\
& + 256a^3e^5 + 3125a^4g^4 + 160a^2gbe^3c + 560a^2gdc^2e^2 \\
& + 1020a^2gbd^2e^2 + 160ag^2b^3ed + 560ag^2d^2cb^2 + 1020ag^2b^2c^2e \\
& - 192b^4ecg^2 + 24ab^2ed^3g + 24abe^2c^3g + 144b^4e^2dg - 6b^3e^2c^2g \\
& + 144dc^2b^3g^2 - 630dac^3bg^2 - 630d^3a^2ceg - 72d^4acbg - 72dac^4eg \\
& - 4d^3c^2b^2g - 1600ag^3cb^3 - 2500a^3g^3be - 50a^2g^2b^2e^2 \\
& - 3750a^3g^3dc + 2000a^2g^3db^2 + 2000a^3g^2ce^2 + 825a^2g^2d^2c^2 \\
& + 2250a^2g^3bc^2 + 2250a^3g^2ed^2 - 900a^2g^2bd^3 - 900a^2g^2c^3e - 36agb^3e^3 \\
& - 1600a^3ge^3d + 16d^3ac^3g - 128d^2b^4g^2 + 16d^4b^3g - 27c^4b^2g^2 \\
& + 108ac^5g^2 + 108a^2d^5g + 256b^5g^3.
\end{aligned}$$

One univariate polynomial is the smallest well-constrained system. We can generalize the definition of discriminant to any well-constrained system of multivariate polynomials. In this chapter we are concerned with systems of polynomials and, in particular, sparse polynomials, in other words polynomials with fixed support, or set of nonzero terms.

A related and equally useful tool is the *resultant*, or eliminant. The solvability of an overconstrained set of multivariate polynomials is equivalent to the vanishing of the resultant, which is again a polynomial in the input coefficients. The resultant generalizes the coefficient matrix of an overconstrained linear system and the Sylvester determinant of two polynomials in a single variable. We shall recall the (sparse) resultant, a fundamental object in sparse (or toric) elimination theory, and we shall connect it to the (sparse) discriminant. Here, sparsity means that only certain monomials in each of the $n+1$ polynomials have nonzero coefficients. Resultants are described in the sequel and their relation to discriminants is explained.

It shall become obvious that computing the (mixed) discriminant is an elimination problem, much akin to resultants. Discriminant computation is NP-hard when the system's dimension varies. There have been several approaches for computing discriminants, based on Gröbner bases or resultants. Recently, in [18], they focused on computing the discriminant of a multivariate polynomial via interpolation, based on [15], which essentially leads to an algorithm for predicting the discriminant's Newton polytope, hence its nonzero terms. This yields a new, efficient, and output-sensitive algorithm which, however, remains to be juxtaposed in practice to earlier approaches.

The rest of this chapter is organized as follows. The next section gives some applications of the discriminant, while Sect. 3 provides some general information about sparse elimination theory and resultants. Section 4 gives a general description of the discriminant and its properties.

2 Applications

In this section, we present some applications of discriminants, in order to motivate their study. Some applications are analyzed in depth, but others are only listed so as to show the breadth of possible applicability.

The main geometric application comes from the fact that discriminants express special geometric configurations, such as tangency between curves in the plane. Consider the following system of two polynomials in two variables:

$$\begin{aligned} f_1 &= ax_1^2 + bx_1x_2 + cx_2^2 + dx_1 + ex_2 + g, \\ f_2 &= hx_1^2 + ix_1x_2 + jx_2^2 + kx_1 + lx_2 + m. \end{aligned}$$

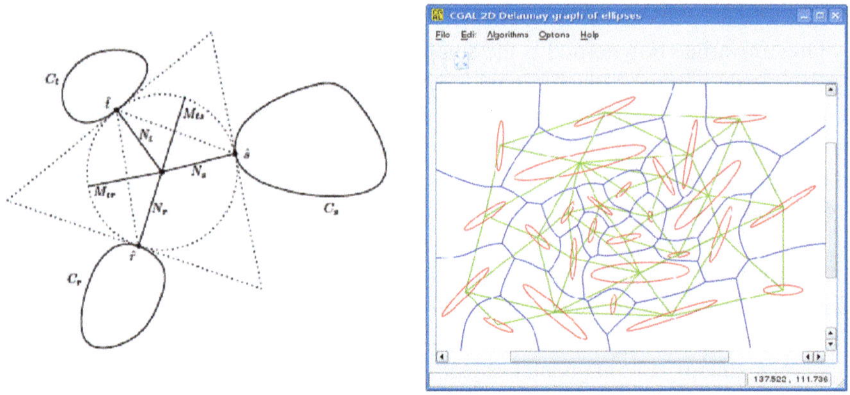

Fig. 1 *Left* Voronoi circle for 3 ellipses. *Right* An example of a Voronoi diagram for non-intersecting ellipses, and the corresponding Delaunay graph. Both figures reproduced from [16]

The condition that the two quadrics f_1, f_2 are tangent is equivalent to the condition that the system's discriminant Δ vanishes, where Δ is of degree 12 and has 3,210 monomials in coefficients $a, b, c, d, e, g, h, i, j, k, l, m$. An interesting remark, which we shall investigate in Sect. 4, is that the discriminant of a well-constrained system can be reduced to that of a single univariate polynomial, albeit of higher degree. In this case, the system's discriminant equals the discriminant of

$$f = ax_1^2x_3 + bx_1x_2x_3 + cx_2^2x_3 + dx_1x_3 + ex_2x_3 + gx_3 + hx_1^2x_4 + ix_1x_2x_4 \\ + jx_2^2x_4 + kx_1x_4 + lx_2x_4 + mx_4.$$

Another geometric application of discriminants is in the computation of the Voronoi diagram of ellipses, or of general smooth convex closed sets, see Fig. 1 (right). Expressing the *Voronoi circle* externally tangent to three given ellipses reduces to a discriminant computation. If the Voronoi circle of center (v_1, v_2) has radius $\sqrt{s}$, then for each of the three ellipses, we consider the resulting discriminant Δ_i expressing tangency between the i-th ellipse and the Voronoi circle, for $i = 1, 2, 3$. We thus get the following 3×3 polynomial system [16]:

$$\Delta_1(v_1, v_2, s) = \Delta_2(v_1, v_2, s) = \Delta_3(v_1, v_2, s) = 0,$$

see Fig. 1 (left). The above system has one common root that specifies the unique Voronoi circle, but has several other roots that correspond to other (complex) tritangent circles to the three ellipses.

For a line to be tangent to two ellipses, the discriminant of the polynomials, expressing tangency between each ellipse and an (unknown) line, should vanish. All the real roots of the discriminant are the values of the parameter that correspond to the tangency points, which in turn allows us to compute the implicit equations of all bitangent lines. There are four bitangents to two disjoint ellipses unless the latter constitute a degenerate situation.

The discriminant is encountered in further geometric applications, for example in the description of the topology of plane curves [23]. In real algebraic geometry, the number of real roots of a real polynomial is constant, when the coefficients vary on each connected component of the zero set (or zero locus) of the (sparse) mixed discriminant, given that for the number of real roots to increase, two complex roots should merge.

Discriminants are thus used in solving systems of polynomial inequalities and in zero-dimensional system solving [19], in root classification and computation of real roots of square systems of sparse polynomials [13], in detecting real roots for n-variate $(n+2)$-monomial polynomials with integer exponents [2], in the description of the topology of real algebraic plane curves [23], and in the determination of cusp points of parallel manipulators [26].

In [17], based on precomputed discriminants, they classify, isolate with rational points, and compare the real roots of polynomials of degree up to four. In [24] they present an algorithm for computing discriminants and prime ideal decomposition in number fields. The algorithm computes the p-valuation of the index of a generating equation $f(x)$ as well; in particular, it determines the discriminant of the number field, once one is able to factorize the discriminant of the defining equation. In [3], a key point is to show that the univariate polynomial below, with rational coefficients, namely:

$$\begin{aligned}x^{17} - 5x^{16} + 12x^{15} - 28x^{14} + 72x^{13} - 132x^{12} + 116x^{11}\\ - 74x^{9} + 90x^{8} - 28x^{7} - 12x^{6} + 24x^{5} - 12x^{4} - 4x^{3} - 3x - 1\end{aligned}$$

has Galois group $SL_2(\mathbb{F}_{16})$. This is achieved by the use of discriminants and their factorization. Furthermore, in [20] they break the Algebraic Surface Cryptosystem (ASC) proposed in 2009. The main idea is to decompose ideals deduced from the ciphertext in order to avoid to solve the section-finding problem. They achieve this by an algorithm that computes and factors resultants.

Another application is an algebraic problem, which arises from considering pairs of differential equations on the plane of the form

$$\dot{x} = P(x, y), \quad \dot{y} = Q(x, y),$$

where P and Q are polynomials in x, y. To find the equilibrium points we have to find the intersections of the curves $P(x, y) = 0 = Q(x, y)$ and also to decide whether they touch at these points, and whether the discriminant Δ vanishes there. Lastly, discriminants can be found in applied physics, e.g. in dark matter search [8]. Moreover, in [22] are established the algebraic conditions for a polynomial to have a pair of minima that have a common horizontal tangent, by computing, among others, a resultant and giving a factorization of the discriminant of a polynomial. This condition is exactly that required by the Maxwell convention of catastrophe theory. The extremal monomials and coefficients of the discriminant have interesting combinatorial descriptions. This has important applications in singularity theory and number theory.

Let us conclude with the following example, which is an application of (sparse) mixed discriminants that concerns the determination of real roots.

Example 1 [14] The Wilkinson polynomial

$$W_{20} = \prod_{i=1}^{20}(x+i) = \sum_{j=0}^{20} c_j x^j$$

is well known for its numerical instability [32]. It has 20 real roots, but the polynomial $W_{20}(x) + 10^{-9}x^{19}$ has only 12 real roots and 4 pairs of complex roots, which do not seem to have small imaginary part, as one of these pairs is approximately equal to $-16.57173899 \pm 0.8833156071i$. On the other hand, if we subtract $10^{-9}x^{19}$ from W_{20} we get a polynomial with 14 real zeros. This unstable behavior could be explained by the fact that the vector of coefficients $(20!, \ldots, 210, 1)$ of W_{20} is very close not only to the variety (set) of ill-posed polynomials, but also very close to a singular point of this variety.

In [14], there are experiments with the following 2-dimensional family of polynomials of degree 20:

$$W(a, b, x) := W_{20}(x) + ax^{19} + bx^{18}.$$

The corresponding discriminant $\Delta(a, b)$ defines a singular curve traced inside the discriminant locus. The singularities of $\Delta(a, b) = 0$ are close to the point $a = b = 0$, i.e., to the vector of coefficients of the *Wilkinson* polynomial. Figure 2 features sample points of $\Delta(a, b) = 0$ inside a small box around the origin, which is the point lying in the intersection of the two coordinate arrows.

Considering the distance, not just to the variety of ill-posed problems, but also to its singular locus would correspond, in the case of conditioning of square $m \times m$ matrices in linear algebra, to consider not only the smallest and greatest singular values, but also the behavior of the intermediate ones.

3 Sparse Elimination Theory

This section introduces sparse (or toric) elimination theory and its main tool, the sparse resultant.

Classical elimination theory and the classical multivariate resultant have a long and rich history that includes such luminaries as Euler, Bézout, Cayley and Macaulay; see [9, 12, 31]. Having been at the crossroads between pure and computational mathematics, it became the victim, in the second quarter of this century, of the

Fig. 2 The figure is by B. Mourrain using software Mathemagix [30], see also [14]

polemic led by the promoters of abstract approaches. Characteristically, the third edition of van der Waerden's *Modern Algebra* has a chapter on elimination theory and resultants that has disappeared from later editions.

Moreover, when the number of variables exceeds four or five, elimination methods lead to matrices which are, of course, too large for hand calculations and quite demanding computationally. However, the advent of modern computers has revived this area. The last decade has seen efficient resultant-based solutions of certain algorithmic as well as applied problems. Some of these problems were impossible to tackle with other methods in real time. These areas include, among others, robotics [5], and geometric modeling [25].

The classical (or projective) resultant of a system of n homogeneous polynomials in n variables vanishes exactly when there exists a common solution in projective space. The sparse (or toric) resultant of $n+1$ polynomials in n variables characterizes solvability over a smaller space which coincides with affine space under certain genericity conditions. Sparse elimination theory concerns the study of resultants and discriminants associated with toric varieties, in other words varieties defined for a given set of support points. This theory has its origin in the work of Gel'fand, Kapranov and Zelevinsky on multivariate hypergeometric functions. Singularities of such functions are discriminants, whereas the denominator of rational hypergeometric functions is a product of resultants, that is, a product of special discriminants.

Let us start with some definitions: $conv(A)$ denotes the convex hull of set A. Volume, denoted by $\mathrm{Vol}(\cdot)$, is always considered normalized with respect to the lattice $\mathbb{Z}^n$, so that a primitive triangle or simplex has volume equal to 1. As usual $Q_1 + Q_2$ denotes the Minkowski sum of Q_1 and Q_2.

The *Newton polytope* $N(f)$ of a nonzero polynomial

$$f = \sum_{a \in A \subset \mathbb{Z}^n} c_a x^a, \quad c_a \neq 0,$$

Fig. 3 The Newton polygons $Q_1' = conv(A_1')$, $Q_1'' = conv(A_1'')$, $Q_2 = conv(A_2)$, and $Q_1 = conv(A_1)$

is the polytope with integer vertices defined as the convex hull of A; the latter is the *support* of f and contains precisely the exponents occurring in f with non-zero coefficients.

Sparsity is measured in geometric terms by the Newton polytope of the polynomial, but what does sparsity mean? The number of nonzero monomials is not necessarily small, but they are modeled by the Newton polytope, thus leading to rich theory, which exploits combinatorial ideas and properties. The main notion, of course, is that a polynomial system is characterized by those monomials in each of the polynomials that have nonzero coefficients. Here is an example.

Example 2 Consider specific polynomials f_1', f_1'', f_2 and their supports, which are full dimensional in $\mathbb{Z}^2$ as follows:

$$A_1' = \{(0,0),(0,1),(1,0)\}, \quad A_1'' = \{(0,0),(0,1),(1,0)\}, \quad A_2 = \{(0,0),(0,1),(1,0)\}.$$

Let $f_1 = f_1' \cdot f_1'$, then its support is $A_1 = A_1' + A_1'' = \{(0,0),(0,1),(1,0),(2,0),(0,2),(1,1)\}$. All Newton polytopes (here, polygons) can be seen in Fig. 3.

Example 3 Consider polynomials f_1', f_1'', f_2 and their supports, as follows:

$$A_1' = \{(0,0),(0,1),(0,2),(1,0)\}, \quad A_1'' = \{(0,0),(0,1),(1,0),(2,0)\},$$

and $A_2 = \{(0,0),(0,1),(0,2),(1,0),(2,0),(2,1),(2,2),(1,2)\}$. If $f_1 = f_1' \cdot f_1''$, then its support is

$$\begin{aligned} A_1 &= A_1' + A_1'' \\ &= \{(0,0),(0,1),(1,0),(2,0),(0,2),(1,1),(2,1),(0,3),(1,2),(2,2),(3,0)\}. \end{aligned}$$

All Newton polytopes (here, polygons) can be seen in Figs. 4 and 5.

The *mixed volume* $MV(Q_1,\ldots,Q_n)$ of n convex polytopes Q_i in $\mathbb{R}^n$ is a classic function in geometry, taking values in $\mathbb{N}$ and generalizing the notion of volume, in the sense that mixed volume equals $n!\,\mathrm{Vol}(Q_1)$, when $Q_1 = \cdots = Q_n$. Mixed volume

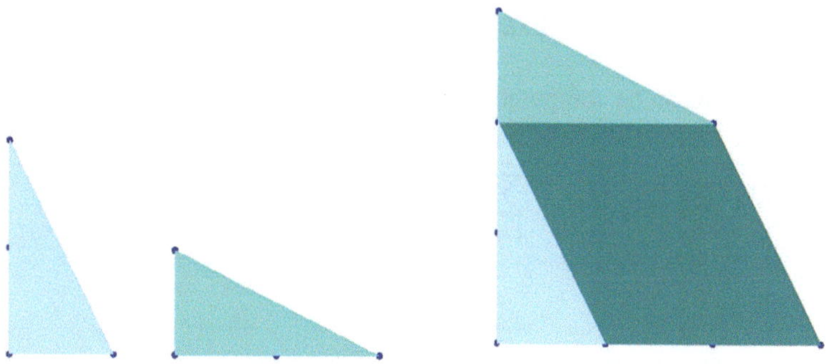

Fig. 4 The Newton polygons $Q_1' = conv(A_1')$, $Q_1'' = conv(A_1'')$, and $Q_1 = conv(A_1)$

Fig. 5 The Newton polygons $Q_2 = conv(A_2)$, and $Q_1 + Q_2 = conv(A_1 + A_2)$

is multilinear with respect to scalar multiplication and Minkowski addition of the Q_i's.

The cornerstone of sparse elimination theory is the following theorem.

Theorem 1 [1] *The mixed volume of the Newton polytopes of polynomials $f_1(x), \ldots, f_n(x)$ in n variables bounds the number of common solutions of $f_1(x) = 0, \ldots, f_n(x) = 0$ in the algebraic torus $(K^*)^n$, where K is an algebraically closed field of the coefficients. If the coefficients of the polynomials are sufficiently generic, then the number of common solutions equals the mixed volume.*

This bound generalizes, to the sparse case, Bézout's classical bound, which is equal to the product of the n polynomials' total degrees, and bounds the number of solutions in n-dimensional complex projective space. For polynomials whose supports are simplices, as in Example 2, the mixed volume and Bézout's bound coincide.

Mixed volume can be computed in terms of Minkowski sum volumes:

$$MV_n(Q_1, \dots, Q_n) = \sum_{k=1}^{n} (-1)^{n-k} \sum_{I \subset \{1,\dots,n\}, |I|=k} \mathrm{Vol}\Big(\sum_{i \in I} Q_i\Big).$$

This implies, for $n = 2$:

$$MV(Q_1, Q_2) = \mathrm{Vol}(Q_1 + Q_2) - \mathrm{Vol}(Q_1) - \mathrm{Vol}(Q_2). \tag{2}$$

In general, this formula does not lead to efficient computation. Instead, an efficient algorithm and implementation has been developed in [6].

3.1 Resultants

The strong interest in multivariate resultants is explained because resultant-based methods have been found to be very efficient for solving certain classes of small and medium-size problems, say of dimension up to 10. For a system of $n + 1$ arbitrary polynomial equations in n variables, it is a polynomial in the coefficients, hence it eliminates n variables. The resultant is defined when all polynomial coefficients are symbolic, but typically used when only some of them are symbolic.

One example is the determinant of the coefficient matrix of $n + 1$ linear polynomials. Unless the coefficients are very particularly chosen, the resultant vanishes for a particular specialization of all coefficients if and only if the given system has a non-trivial solution.

Another example is the Sylvester resultant, namely for $n = 1$. Then, the resultant equals the determinant of Sylvester's matrix. For generic polynomials $f_1(x), f_2(x)$ of degrees one and two, respectively, Sylvester's matrix S is as follows:

$$\left\{ \begin{array}{ll} f_1(x) = & a_1 x + a_0 \\ f_2(x) = & b_2 x^2 + b_1 x + b_0 \end{array} \right\} \quad \text{and} \quad S = \begin{bmatrix} a_1 & a_0 & 0 \\ 0 & a_1 & a_0 \\ b_2 & b_1 & b_0 \end{bmatrix}. \tag{3}$$

The resultant equals $\det S = a_1^2 b_0 + a_0^2 b_2 - a_0 a_1 b_1$. Thus S is an instance of a resultant matrix, in other words a matrix whose determinant yields the resultant. The principal merit of resultant matrices is that they reduce the solution of a non-linear system to a matrix problem, where we can use an arsenal of numeric linear algebra techniques and software, see e.g. [9, 12]. By construction, the existence of common solutions implies a decrease of matrix rank.

In most applications, we deal with *well-constrained* systems, namely systems of k polynomials in k unknowns. To obtain an overconstrained system, for which the resultant is defined, we should either add an extra polynomial or "hide" a variable in the coefficient field [9, 12, 31].

We now formally define the resultant polynomial of an overconstrained system.

Definition 1 The *resultant* $R(f_0, \ldots, f_n)$ of $n+1$ polynomials $f_0, \ldots, f_n$ in n variables is an irreducible polynomial in the coefficients of $f_0, \ldots, f_n$, which vanishes whenever $f_0, \ldots, f_n$ have a common root.

The *sparse resultant* has an analogous definition when the f_i are specified by their supports $A_i \subseteq \mathbb{Z}^n$. Formally, the *sparse resultant* of $f_0, \ldots, f_n$, where each f_i has support A_i, is an irreducible polynomial with integer coefficients over the coefficients of the f_i such that it vanishes precisely when the system $f_0 = f_1 = \cdots = f_n = 0$ has a solution in $(\mathbb{C}^*)^n$.

The Newton polytope of the (sparse) resultant is called the resultant polytope, and it can be effectively computed by the algorithm in [15], see also [28].

4 Discriminants

In this section we introduce discriminants of well constrained systems, provide some definitions and overview relevant properties in Sect. 4.1.

Gel'fand et al. [21] established the following definition, which we shall formally state later: The (sparse) *mixed discriminant* $\Delta(f_1, \ldots, f_n)$ of n polynomials in n variables is the irreducible polynomial in the coefficients of the f_i which vanishes whenever the system $f_1 = \cdots = f_n = 0$ has a multiple root or, equivalently, a root which is not simple. Consider a system of two polynomials in two variables:

$$\begin{aligned} f_1(x_1, x_2) &= a_0 + a_1x_1 + a_2x_1^2 + a_3x_2 + a_4x_2^2 + a_5x_1x_2, \\ f_2(x_1, x_2) &= b_0 + b_1x_1 + b_2x_2. \end{aligned}$$

Their discriminant shows whether a common root is singular.

Let us formalize the definition of (sparse) discriminants. We consider n (finite) lattice configurations $A_1, \ldots, A_n$ in $\mathbb{Z}^n$ and we denote by $Q_1, \ldots, Q_n$ their respective convex hulls. Let $f_1, \ldots, f_n$ be Laurent polynomials with support $A_1, \ldots, A_n$ respectively:

$$f_i(x) = \sum_{\alpha \in A_i} c_{i,\alpha} x^\alpha, \quad i = 1\ldots, n. \tag{4}$$

We define the *discriminantal variety* to be the closure of the locus of coefficients $c_{i,a}$ for which the associated system of n polynomial equations in n unknowns $x = (x_1, \ldots, x_n)$, over an algebraically closed field K, namely:

$$f_1(x) = \cdots = f_n(x) = 0, \tag{5}$$

has a non-degenerate multiple root.

Before the general definition we present the simplest case.

Example 4 Given a generic univariate polynomial of degree d,

$$P(z) = a_0 + a_1 z + \cdots + a_d z^d, \quad a_d \neq 0,$$

there exists an irreducible polynomial $\Delta(P) = \Delta(a_0, \ldots, a_d) \in \mathbb{Z}[a_0, \ldots, a_d]$, unique up to sign, called the *discriminant*, which verifies $\Delta(a_0, \ldots, a_n) \neq 0$ if and only if all roots of P are simple for any specialization of the coefficients in $\mathbb{C}$, with $a_d \neq 0$. Thus $\Delta(a_0, \ldots, a_n) = 0$ if and only if there exists $z \in \mathbb{C}$ with $P(z) = P'(z) = 0$. In fact, the corresponding Sylvester resultant $R(P, P')$ equals $a_d \Delta(P)$.

This discriminant is an instance of both A-discriminant and mixed discriminant and which are defined below.

Geometrically, the discriminant hypersurface

$$\{a = (a_0, \ldots, a_d) \in \mathbb{C}^{d+1} : \Delta(a) = 0\}$$

is the projection over the first $d+1$ coordinates of the intersection of the hypersurfaces $\{(a, z) \in \mathbb{C}^{d+2} : a_0 + a_1 z + \cdots + a_d z^d = 0\}$ and $\{(a, z) \in \mathbb{C}^{d+2} : a_1 + 2a_2 z + \cdots + d a_d z^{d-1} = 0\}$, in other words the variable z is eliminated.

An isolated solution $u \in (K^*)^n$ is a *nondegenerate multiple root* if the n gradient vectors

$$\left(\frac{\partial f_i}{\partial x_1}(u), \ldots, \frac{\partial f_i}{\partial x_n}(u) \right)$$

are linearly dependent, but any $n - 1$ of them are linearly independent.

We now give the definition of (sparse) A-discriminant from [21]. It is related to some support set A, thus capturing the sparse structure of the data.

Definition 2 Consider the polynomial defined from system (4) for $n = 1$. We denote by Δ_A the (sparse) *A-discriminant*, which is the unique (up to sign) irreducible polynomial with integer coefficients, in the parameter coefficients $c_{i,a}$, where we follow the previous notation. Δ_A vanishes whenever the hypersurface is not smooth. Otherwise we refer to set A as a *defective* support, and set $\Delta_A = 1$.

Let $A_1, \ldots, A_n$ be pointsets in $\mathbb{Z}^n$, as specified for system (4). We define the (sparse) mixed discriminant from [7], which captures the structure in a given well-constrained polynomial system.

Definition 3 If the discriminantal variety is a hypersurface, we define the (sparse) *mixed discriminant* of system (5) to be the unique up to sign irreducible polynomial $\Delta_{A_1,\ldots,A_n}$ with integer coefficients in the unknown parameters $c_{i,a}$, which defines this hypersurface. Otherwise, we say that the system is defective and set $\Delta_{A_1,\ldots,A_n} = 1$.

We now relate the previous two notions with an important construction in algebraic combinatorics. Let $A_1, \dots, A_n$ be supports in $\mathbb{Z}^n$, defining Laurent polynomials, then A shall be specified to be the corresponding Cayley matrix. This matrix is defined to ave $2n$ rows and $m = \sum_{i=1}^{n} |A_i|$ columns. We introduce n new variables $y_1, \dots, y_n$ in order to encode the system $f_1 = \dots = f_n = 0$ in a single polynomial with support in A. This is known as the *Cayley trick* and yields polynomial

$$\phi(x, y) = y_1 f_1(x) + \dots + y_n f_n(x).$$

The Cayley matrix is:

$$A = \begin{pmatrix} 1 & 0 & \dots & 0 \\ 0 & 1 & \dots & 0 \\ \dots & \dots & \dots & \dots \\ 0 & 0 & \dots & 0 \\ A_1 & A_2 & \dots & A_n \end{pmatrix}.$$

By [7, Theorem 2.1], the mixed discriminant $\Delta_{A_1,\dots,A_n}$ equals the A-discriminant of the associated Cayley matrix whenever $\Delta_A \neq 1$. Let us give an example of the relation between A-discriminant and mixed discriminant.

Example 5 Consider two planar configurations $A_1 = \{(6,0), (0,3), (0,1)\}$, and $A_2 = \{(0,6), (3,0), (1,0)\}$. These supports correspond to the following family of polynomials:

$$h_1(x, y) = c_{11}x^6 + c_{12}y^3 + c_{13}y,$$
$$h_2(x, y) = c_{21}y^6 + c_{22}x^3 + c_{23}x.$$

We introduce two new variables a, b in order to encode the system $h_1 = h_2 = 0$ in a single polynomial, namely

$$\phi(x, y, a, b) = ah_1(x, y) + bh_2(x, y).$$

Then, A is the Cayley matrix associated to the supports A_1, A_2:

$$A = \begin{pmatrix} 1 & 1 & 1 & 0 & 0 & 0 \\ 0 & 0 & 0 & 1 & 1 & 1 \\ 6 & 0 & 0 & 0 & 3 & 1 \\ 0 & 3 & 1 & 6 & 0 & 0 \end{pmatrix},$$

and the A-discriminant $\Delta_A(c) = \Delta_A(c_{11}, \dots, c_{23})$ is the mixed discriminant of h_1, h_2. Now $\Delta_A(c) = 0$ whenever there exists a common zero $(x, y) \in K^2$, making both h_1, h_2 vanish, which is not simple.

It turns out that $\Delta_A(c)$ is a polynomial of degree 90 in c, with 58 monomials and huge integer coefficients.

4.1 Properties

In this section we review existing work on the discriminant degree, and on multiplicativity formulae for sparse resultants and discriminants.

We start with the necessary notation for a theorem on the degree of the mixed discriminant of two polynomials with fixed supports. A subset $F \subseteq A$ is called *face* of A, if F is the intersection of A with a face of the polytope $conv(A)$. We shall write $R(f_1, \ldots, f_n)$ and $\Delta(f_1, \ldots, f_n)$ without subscripts to imply $R_{A_1,\ldots,A_n}(f_1, \ldots, f_n)$ and $\Delta_{A_1,\ldots,A_n}(f_1, \ldots, f_n)$. Let π denote the projection to $\mathbb{R}A/\mathbb{R}F$, where F is a face of A. We set

$$u(F, A) := \mathrm{Vol}\Big(conv(\pi(A)) - conv(\pi(A - F))\Big).$$

If e_1 and e_2 are parallel edges in Q_1 and Q_2 with same orientation, that is, same inward normal direction, then we call them *strongly parallel*. Let E_i denote the set of edges of A_i and set

$$P := \{(e_1, e_2) \in E_1 \times E_2 : e_1 \text{ is strongly parallel to } e_2\}.$$

We write $l(e)$ for the *normalized length* of an edge e with respect to the lattice $\mathbb{Z}^n$. If $v \in \mathrm{Vert}(A_2)$, where $\mathrm{Vert}(A_2)$ are the vertices of A_2, we define its *mixed multiplicity* as follows:

$$mm(v) := MV(Q_1, Q_2) - MV(conv(A_2 - \{v\}), Q_1).$$

Let us introduce polynomial

$$\tilde{\Delta}_{A_1,\ldots,A_n} = \Delta_{A_1,\ldots,A_n}^{i(A_1,\ldots,A_n)}$$

defined as the power of the mixed discriminant $\Delta_{A_1,\ldots,A_n}$ raised to the index

$$i(A_1, \ldots, A_n) = [\mathbb{Z}^n : \mathbb{Z}A_1 + \cdots + \mathbb{Z}A_n].$$

The latter stands for the index of lattice $\mathbb{Z}A_1 + \cdots + \mathbb{Z}A_{n+1}$ in $\mathbb{Z}^n$, as a subgroup. In general, this index equals 1. Let the discriminant degree in the coefficients of the i-th input polynomial be denoted by

$$\delta_i = deg_{A_i}(\tilde{\Delta}_{A_i,A_j}), \quad i = 1, 2.$$

Then, the following theorem holds.

Theorem 2 [7] *Let A_1 and A_2 be full-dimensional supports in $\mathbb{Z}^n$. Then δ_i equals*

$$2 \cdot Vol(Q_j) + 2 \cdot MV(Q_i, Q_j) - \sum_{(e_i,e_j)\in P} \min\{u(e_i, A_i), u(e_j, A_j)\} l(e_j) - \sum_{v \in Vert(A_i)} mm(v),$$

where $i = 1, 2$ and $j = 2, 1$, in other words $\{i, j\} = \{1, 2\}$.

An explicit degree formula for the special cases of plane curves is also presented in [7, Corollary 3.15]. We correct this formula. The degree of Δ_{A_1,A_2} can be computed as follows:

$$\delta_1 = area(Q_1 + Q_2) + area(Q_1) - perimeter(Q_2),$$

$$\delta_2 = area(Q_1 + Q_2) + area(Q_2) - perimeter(Q_1),$$

where $Q_i = conv(A_i)$, $i = 1, 2$, $Q_1 + Q_2$ is their Minkowski sum. The area (like volume above) is normalized, so that a primitive triangle has area 1 and the perimeter of Q_i is the cardinality of $\partial Q_i \cap \mathbb{Z}^2$.

Computing resultants and discriminants is usually a computationally hard task. However, if one polynomial factors as $f_1 = f_1' \cdot f_1''$, both resultant and discriminant factors, and we thus obtain a multiplicativity formula. This significantly simplifies the corresponding computation.

Let us recall the case when one polynomial factors as a product of two polynomials, in the case of resultants. The multiplicativity formula for sparse resultants can be found in [27, Proposition 7.1], see also [10]. Consider polynomials $f_1, \dots, f_{n+1}$ in variables $x_1, \dots, x_n$ and let $f_1 = f_1' \cdot f_1''$ be a product of two polynomials, where all relevant supports are $A_1, A_1', A_1'', A_2, \dots, A_{n+1} \subseteq \mathbb{Z}^n$ respectively. Then,

$$R(f_1'f_1'', f_2, \dots, f_{n+1}) = R(f_1', f_2, \dots, f_{n+1}) \cdot R(f_1'', f_2, \dots, f_{n+1}). \quad (6)$$

We now pass to discriminants. The multiplicativity property of the discriminant in the case of (dense) homogeneous polynomials was already known to Sylvester [29], and has been generalized by Busé and Jouanolou [4]. They prove that for any n, when all A_i's correspond to (the lattice points in) a dilate of the standard simplex and $A_i = A_1' + A_1''$ is the sum of two dilates of the simplex then, given polynomials $f_1', f_1'', f_2, \dots, f_n$ with corresponding supports $A_1', A_1'', A_2, \dots, A_n$,

$$\Delta(f_1, \dots, f_n) = \Delta(f_1', \dots, f_n) \cdot \Delta(f_1'', \dots, f_n) \cdot R(f_1', f_1'', \dots, f_n)^2.$$

Recall that the discriminant of $n-1$ homogeneous polynomials of degree 1 equals 1 by convention.

Theorem 3 [4, 29] *Let $f_1', f_1'', f_2, \dots, f_{n-1}$ be n homogeneous polynomials in $R[x_1, \dots, x_n]$, of degrees $d_1', d_1'', d_2, \dots, d_{n-1} \geq 1$, respectively.*
Then $\Delta(f_1'f_1'', f_2, \dots, f_{n-1})$ factors as follows:

$$(-1)^{d_1'd_1''d_2\dots d_{n-1}} \cdot \Delta(f_1', f_2, \dots, f_{n-1}) \cdot \Delta(f_1'', f_2, \dots, f_{n-1}) \cdot R(f_1', f_1'', f_2, \dots, f_{n-1})^2.$$

Our current work focuses on multiplicativity formulas for the mixed discriminant in the case $n = 2$, with fixed supports, always within the realm of sparse elimination theory [11], aiming at efficient algorithms. It turns out that a key issue is to understand the relation between the mixed discriminant of two bivariate Laurent polynomials, where one factors, and the sparse resultant of those three bivariate polynomials.

Acknowledgments We thank Alicia Dickenstein for useful discussions. I. Z. Emiris was partially supported by Marie-Curie Initial Training Network "SAGA" (ShApes, Geometry, Algebra), FP7-PEOPLE contract PITN-GA-2008-214584. A. Karasoulou's research leading to these results has received funding from the European Union (European Social Fund—ESF) and Greek national funds through the Operational Program "Education and Lifelong Learning" of the National Strategic Reference Framework (NSRF)—Research Funding Program "ARISTEIA", project "ESPRESSO: Exploiting Structure in Polynomial Equation and System Solving in Physical Modeling".

References

1. Bernstein, D.N.: The number of roots of a system of equations. Funct. Anal. Appl. **9**(2), 183–185 (1975)
2. Bihan, F., Rojas, J.M., Stella, C.E.: Faster real feasibility via circuit discriminants. In: Proceedings of ACM ISSAC, Seoul, Korea (2009)
3. Bosman, J.: A polynomial with Galois group $SL_2(\mathbb{F}_{16})$. LMS J. Comp. & Math. **10**, 378–388 (2007)
4. Busé, L., Jouanolou, J.: A computational approach to the discriminant of homogeneous polynomials. Technical Report. http://hal.inria.fr/hal-00747930/en/ (2012)
5. Canny, J.F.: The Complexity of Robot Motion Planning. M.I.T. Press, Cambridge (1988)
6. Canny, J.F., Emiris, I.Z.: Efficient incremental algorithms for the sparse resultant and the mixed volume. J. Symb. Comput. **20**(2), 117–149 (1995)
7. Cattani, E., Cueto, M.A., Dickenstein, A., Di Rocco, S., Sturmfels, B.: Mixed discriminants. Math. Z. **274**, 761–778 (2013)
8. Cheng, H., Han, Z.: Minimal kinematic constraints and MT2. J. High-Energy Phys. **2008**(12), 063 (2008)
9. Cox, D., Little, J., O'Shea, D.: Using Algebraic Geometry. Springer, New York (1997)
10. D'Andrea, C., Sombra, M.: A Poisson Formula for the Sparse Resultant. Preprint, arXiv:1310.6617 (2013)
11. Dickenstein, A., Emiris, I.Z., Karasoulou, A.: SAGA: Advances in shapes, algebra, and geometry. Plane Mixed Discriminants and Toric Jacobians (2014)
12. Dickenstein, A., Emiris, I.Z. (eds.): Solving Polynomial Equations, Algorithms and Computation in Mathematics 14. Springer, Heidelberg (2005)
13. Dickenstein, A., Rojas, J.M., Rusek, K., Shih, J.: Extremal real algebraic geometry and a-discriminants. Moscow Math. J. **7**, 425–452 (2007)
14. Dickenstein, A.: A world of binomials, foundations of computational mathematics (hong kong 2008). Lond. Math. Soc. Lec. Note Ser. **363**, 42–66 (2009)
15. Emiris, I.Z., Fisikopoulos, V., Konaxis, C., Penaranda, L.: An output-sensitive algorithm for computing projections of resultant polytopes. In: Proceedings of ACM Symposium Computational Geometry, pp. 179–188 (2012). To appear in Intern. J. Comput. Geom. Appl.
16. Emiris, I.Z., Tsigaridas, E., Tzoumas, G.: Exact Voronoi diagram of smooth convex pseudo-circles: General predicates, and implementation for ellipses. J. Comput. Aided. Geom. Des. **30**(8), 760–777 (2013)
17. Emiris, I.Z., Tsigaridas, E.P.: Real algebraic numbers and polynomial systems of small degree. Theor. Comput. Sci. **409**(2), 186–199 (2008)
18. Emiris, I.Z., Kalinka, T., Konaxis, C., Luu Ba, T.: Sparse implicitization by interpolation:characterizing non-exactness and an application to computing discriminants. J. Comput. Aided Des. **45**, 252–261 (2013)

19. Faugére, J.-C., Moroz, G., Rouillier, F., Safey El Din, M.: Classification of the perspective-three-point problem: Discriminant variety and real solving polynomial systems of inequalities. In: Proceedings of ACM ISSAC, Hagenberg (Austria), pp. 79–86 (2008)
20. Faugére, J.-C., Spaenlehauer, P.-J.: Algebraic cryptanalysis of the PKC 2009 algebraic surface cryptosystem. In: Public Key Cryptography, pp. 35–52 (2010)
21. Gel'fand, I.M., Kapranov, M.M., Zelevinsky, A.V.: Discriminants, Resultants and Multidimensional Determinants, Mathematics: Theory & Applications. Birkhäuser, Boston (1994)
22. Godwin, A.N.: The precise determination of maxwell sets for cuspoid catastrophes. Int. J. Math. Edu. Sci. & Technol. **15**(2), 167–182 (2006)
23. González-Vega, L., Nacula, I.: Efficient topology determination of implicitly defined algebraic plane curves. Comput. Aided Geom. Des. **19**(9), 719–743 (2002)
24. Guárdia, J., Montes, J., Nart, E.: Higher newton polygons in the computation of discriminants and prime ideal decomposition in number fields. J. de Théorie des Nombres de Bordeaux **23**, 667–696 (2011)
25. Manocha, D., Demmel, J.: Algorithms for intersecting parametric and algebraic curves ii: multiple intersections. Graph. Models Image Proc. **57**(2), 81–100 (1995)
26. Moroz, G., Rouiller, F., Chablat, D., Wenger, P.: On the determination of cusp points of 3-rpr parallel manipulators. Mech. & Mach. Theor. **45**(11), 1555–1567 (2010)
27. Pedersen, P., Sturmfels, B.: Product formulas for resultants and chow forms. Math. Z. **214**(1), 377–396 (1993)
28. Sturmfels, B.: On the Newton polytope of the resultant. J. Algebraic Combin. **3**(2), 207–236 (1994)
29. Sylvester, J.J.: Sur l'extension de la théorie des résultants algébriques. Comptes Rendus de l'Académie des Sci. **LVIII**, 1074–1079 (1864)
30. van der Hoeven, J., Lecerf, G., Mourain, B.: Mathemagix. http://www.mathemagix.org/ (2002)
31. van der Waerden, B.L.: Modern Algebra, 3rd edn. F. Ungar Publishing Co., New York (1950)
32. Wilkinson, J.H.: Rounding Errors in Algebraic Processes. Dover, Mineola (1994)

Parametrization of the Bisector of Two Low Degree Surfaces

Ibrahim Adamou, Mario Fioravanti and Laureano Gonzalez-Vega

Abstract The bisectors are geometric constructions with different applications in tool path generation, motion planning, NC-milling, etc. We present a new approach to determine an algebraic representation (parameterization or implicit equation) of the bisector surface of two given low degree parametric surfaces. The method uses the so-called generalized Cramer rules, and suitable elimination steps. The new introduced approach allows to easily obtain parameterizations of the plane-quadric, plane-torus, circular cylinder-quadric, circular cylinder-torus, cylinder–cylinder, cylinder-cone and cone–cone bisectors, which are rational in most cases. In the remaining cases the parametrization involves one square root, which is well suited for a good approximation of the bisector.

Keywords Bisector surface · Parametrization · Quadric · Cramer's rule

Mathematics Subject Classification (2010) 65D17 · 68U07

1 Introduction

The (untrimmed) bisector of two smooth surfaces is the set of centers of spheres which are tangent to both surfaces (see Fig. 1). The purpose of this work is to present a new approach to compute the parametrization (rational or non-rational) for

I. Adamou (✉) · M. Fioravanti · L. Gonzalez-Vega
Departamento de Matematicas, Estadística y Computación, Universidad de Cantabria, Santander, Spain
e-mail: ibrahim.adamou@unican.es

M. Fioravanti
e-mail: fioravam@unican.es

L. Gonzalez-Vega
e-mail: laureano.gonzalez@unican.es

R. De Amicis and G. Conti (eds.), *Future Vision and Trends on Shapes, Geometry and Algebra*, Springer Proceedings in Mathematics & Statistics 84,
DOI: 10.1007/978-1-4471-6461-6_5,

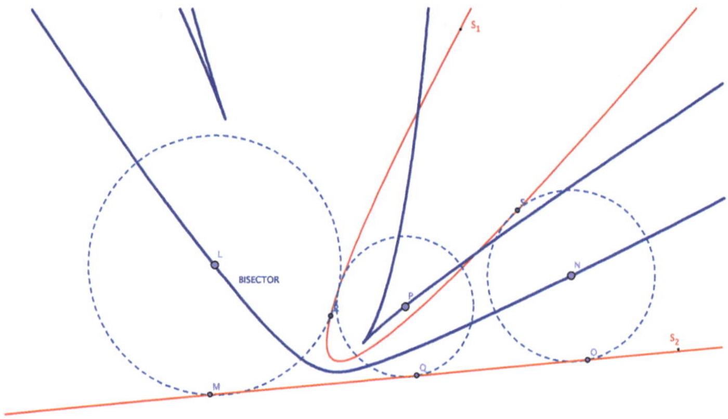

Fig. 1 Schematic illustration of the bisector definition

the bisector of two low degree rational surfaces, given by their parameterizations. We will give special attention to the case where one of the surfaces is a plane or a cylinder. The bisector surfaces are used in several areas of applications, such as tool path generation, motion planning, NC-milling, medial axis transform, Voronoi diagram computation, etc.

Most of the known methods for computing the exact description of the bisectors are devised only for those bisectors possessing rational parameterizations. Various approaches are appropriately used for very special cases to determine a rational representation for the bisector (see for example [3]). In other cases, symmetry considerations reduce the bisector computation to the following cases: point-line, point-surface, or curve-surface [4, 5], where the bisector is a rational surface.

A PN-surface is a surface admitting a parametrization such that the norm of the normal vector is rational. Using Laguerre geometry, Martin Peternell [10, 11] has shown the rationality of the bisector between: plane and PN-surfaces, two PN developable surfaces, two canal surfaces and some other cases.

In the case of non-rationality of the bisector, the exact description is scarce, the implicit representation is of very high degree and often it is impractical to compute. Thus, B-spline and other kinds of approximation have been proposed (see for instance [6]).

Using the so-called generalized Cramer rule [1, 2] and suitable elimination technics we present a new algorithmic approach for computing an algebraic parametrization (rational or non-rational) for the bisector surface of two low degree rational surfaces. Some of the obtained results coincide with those mentioned above in the rational case. The method is well-suited for approximation purposes, which is of special interest in the non-rational case. The new introduced approach allows to easily obtain parameterizations of the plane-quadric, plane-torus, circular cylinder-quadric, circular cylinder-torus, cylinder–cylinder, cylinder-cone and cone–cone bisectors, which are rational in most cases. In the remaining cases, the parametrization involves square roots.

Table 1 Plane-quadric/torus bisector

Quadric/torus	Parametrization	Max. deg. of implicit
Parabolic cylinder	Rational	6
Circular cylinder	Rational	4
Elliptic cylinder	Non rational	8
Hyperbolic cylinder	Non rational	8
Circular cone	Rational	4
Elliptic cone	Non rational	8
Sphere	Rational	4
Ellipsoid	Rational	12
Elliptic paraboloid	Rational	10
Hyperbolic paraboloid	Rational	10
Hyperboloid one-sheet	Rational	12
Hyperboloid two-sheets	Rational	12
Torus	Rational	8

For the case of plane and quadric, or plane and torus, the implicit equation of the bisector can be easily computed, either from the equations defining the bisector, or from the computed parametrization. The results we obtained for the parametrization, as well as for the implicit equation, assuming that the quadric is provided with its PN-parameterization (if possible), are summarized in Table 1. They coincide with those proved, using a different approach, by Peternell [10].

2 Equations of the Bisector

Let S_1 and S_2 be two regular rational surfaces, and $S_1(s,t)$ and $S_2(u,v)$ their parameterizations, respectively. A point $\mathbf{B} = \big(X, Y, Z\big) \in \mathbb{R}^3$ is in the bisector of the surfaces S_1 and S_2 if it satisfies the following system of equations (see [9]):

- the point $\mathbf{B}$ is in the normal lines of S_1 and S_2, at $S_1(s,t)$ and $S_2(u,v)$, respectively:

$$\begin{aligned} \langle (X,Y,Z) - S_1(s,t), \partial_s S_1(s,t)\rangle &= 0,\\ \langle (X,Y,Z) - S_1(s,t), \partial_t S_1(s,t)\rangle &= 0,\\ \langle (X,Y,Z) - S_2(u,v), \partial_u S_2(u,v)\rangle &= 0,\\ \langle (X,Y,Z) - S_2(u,v), \partial_v S_2(u,v)\rangle &= 0, \end{aligned} \tag{1}$$

- the point $\mathbf{B}$ is at equal distance from $S_1(s,t)$ and $S_2(u,v)$:

$$\langle (X,Y,Z), 2(S_2(u,v) - S_1(s,t))\rangle + \|S_1(s,t)\|^2 - \|S_2(u,v)\|^2 = 0. \tag{2}$$

The Eq. 1 can be written in matrix form as follows

$$\mathbf{A}\,\mathbf{B}^T = \mathbf{R}, \tag{3}$$

where

$$\mathbf{A} = \begin{bmatrix} \partial_s S_{1x} & \partial_s S_{1y} & \partial_s S_{1z} \\ \partial_t S_{1x} & \partial_t S_{1y} & \partial_t S_{1z} \\ \partial_u S_{2x} & \partial_u S_{2y} & \partial_u S_{2z} \\ \partial_v S_{2x} & \partial_v S_{2y} & \partial_v S_{2z} \end{bmatrix}, \quad \mathbf{R} = \begin{bmatrix} \langle S_1, \partial_s S_1 \rangle \\ \langle S_1, \partial_t S_1 \rangle \\ \langle S_2, \partial_u S_2 \rangle \\ \langle S_2, \partial_v S_2 \rangle \end{bmatrix}.$$

3 General Approach for Computing the Bisector Parametrization

Our goal is to compute a parametrization of the bisector of $S_1(s,t)$ and $S_2(u,v)$ in terms of two parameters chosen from u, v, s and t. If for some values of the parameters u, v, s, t, the linear system (3) of unknown $\mathbf{B}$ admits a solution, then $\mathrm{rank}(|\mathbf{A}, \mathbf{R}|) \leqslant 3$, and thus $\det(|\mathbf{A}, \mathbf{R}|) = 0$.

Our approach consists in:

- First solve the system (3) for $\mathbf{B}$ in terms of u, v, s, t, using the generalized inverse matrix, $\mathbf{A}^+$, of the matrix $\mathbf{A}$. Substituting $\mathbf{B}(u,v,s,t)$ in (2) we obtain the equation:

$$F(u,v,s,t) = \langle \mathbf{B}, 2(S_2(u,v) - S_1(s,t)) \rangle + \|S_1(s,t)\|^2 - \|S_2(u,v)\|^2. \tag{4}$$

- Then, eliminate two of the four parameters u, v, s and t from

$$G(u,v,s,t) = \det(|\mathbf{A}, \mathbf{R}|) = 0, \quad F(u,v,s,t) = 0, \tag{5}$$

  in the possible and suitable way as follows:

  - Express one of the parameters, say t, in terms of u, v and s from $G(u,v,s,t) = 0$, namely $t = t(u,v,s)$, and substitute in $F(u,v,s,t) = 0$ to obtain: $F_0(u,v,s) = F(u,v,t(u,v,s)) = 0$.
  - Then express s in terms of u and v from $F_0(u,v,s) = 0$, namely $s = s(u,v)$. There might be more than one solution for s (see Example 1).

- Finally, substitute successively t by $t(u,v,s)$ and s by $s(u,v)$, in $\mathbf{B}(u,v,s,t)$, for each solution. We obtain parameterizations of the bisector components, of the form:

$$\mathbf{B}(u,v) = [x(u,v), y(u,v), z(u,v)]^T,$$

  where $x(u,v)$, $y(u,v)$, $z(u,v)$ are in general non-rational.

The process is summarized in the following algorithm:

Algorithm 1

`Input:` $\mathbf{S_1}(s,t)$, $\mathbf{S_2}(u,v)$, the parameterizations of two rational surfaces.
`Output:` $\mathbf{B}_1(u,v),\ldots,\mathbf{B}_m(u,v)$, the algebraic parameterizations of the components of the bisector of the surfaces S_1 and S_2.

1. Compute the Moore-Penrose generalized inverse of $\mathbf{A}$, $\mathbf{A}^+$ (see [1]):

$$\mathbf{A}^+ = -a_k^{-1}\mathbf{A}^T\left[(\mathbf{A}\mathbf{A}^T)^{k-1} + a_1(\mathbf{A}\mathbf{A}^T)^{k-2} + \cdots + a_{k-1}I_n\right],$$

where a_i and k are such that $\det(wI_n - \mathbf{A}\mathbf{A}^T) = a_0w^n + a_1w^{n-1} + \cdots + a_{n-1}w + a_n$, with $a_0 = 1$; ($k \neq 0$ is the largest index such that $a_k \neq 0$) or ($k = 0$ and $\mathbf{A}^+ = 0$).
2. Solve the system (3) for $\mathbf{B}$ in terms of u, v, s and t:

$$\mathbf{B}^T(u,v,s,t) = \mathbf{A}^+\mathbf{R}.$$

3. Find the expression of two of the four parameters (say s and t, for example) in terms of the other two, from $G(u,v,s,t) = 0,\ F(u,v,s,t) = 0$:
   - $t = t(u,v,s)$
   - $s_i = s_i(u,v), i = 1,\ldots,m$ (we may obtain several solutions of the equations)
4. Substituting successively t by $t(u,v,s)$ and s by $s_i(u,v)$ in $\mathbf{B}(u,v,s,t)$ we obtain the parameterizations

$$\mathbf{B}_i(u,v) = [x_i(u,v), y_i(u,v), z_i(u,v)]^T,\ i = 1,\ldots,m.$$

Using this algorithm, we can deal with the specific cases of plane-quadric, plane-torus, circular cylinder-quadric, circular cylinder-torus, cylinder–cylinder, cylinder-cone and cone–cone bisectors. The following example presents the application of the algorithm to a particular case.

Example 1 Let

$$S_1(u,v) = \begin{bmatrix} \dfrac{4u}{1+u^2+v^2} \\ \dfrac{4v}{1+u^2+v^2} \\ \dfrac{2(1-u^2-v^2)}{1+u^2+v^2} + 1 \end{bmatrix}, \quad S_2(s,t) = \begin{bmatrix} \dfrac{6s}{1+s^2+t^2} \\ \dfrac{6t}{1+s^2+t^2} \\ \dfrac{3(1-s^2-t^2)}{1+s^2+t^2} - 5 \end{bmatrix}$$

be two non-intersecting spheres.

The corresponding system for their bisector is given by:

$$\begin{bmatrix} \frac{-6(-1+s^2-t^2)}{\rho(s,t)^2} & \frac{-12st}{\rho(s,t)^2} & \frac{-12s}{\rho(s,t)^2} \\ \frac{-12st}{\rho(s,t)^2} & \frac{6(1+s^2-t^2)}{\rho(s,t)^2} & \frac{-12t}{\rho(s,t)^2} \\ \frac{-4(-1+u^2-v^2)}{\rho(u,v)^2} & \frac{-8uv}{\rho(u,v)^2} & \frac{-8u}{\rho(u,v)^2} \\ \frac{-8uv}{\rho(u,v)^2} & \frac{4(1+u^2-v^2)}{\rho(u,v)^2} & \frac{-8v}{\rho(u,v)^2} \end{bmatrix} \begin{bmatrix} X \\ Y \\ Z \end{bmatrix} = \begin{bmatrix} \frac{60s}{\rho(s,t)^2} \\ \frac{60t}{\rho(s,t)^2} \\ \frac{-8u}{\rho(u,v)^2} \\ \frac{-8v}{\rho(u,v)^2} \end{bmatrix},$$

where $\rho(\mu, \nu) = 1 + \mu^2 + \nu^2$.

The generalized inverse matrix $\mathbf{A}^+$ of matrix $\mathbf{A}$ and the bisector point $\mathbf{B}(u, v, s, t)$ in terms u, v, s, t are very large. The associated matrix $|\mathbf{A}, \mathbf{R}|$ is given by:

$$|\mathbf{A}, \mathbf{R}| = \begin{bmatrix} \frac{-6(-1+s^2-t^2)}{\rho(s,t)^2} & \frac{-12st}{\rho(s,t)^2} & \frac{-12s}{\rho(s,t)^2} & \frac{60s}{\rho(s,t)^2} \\ \frac{-12st}{\rho(s,t)^2} & \frac{6(1+s^2-t^2)}{\rho(s,t)^2} & \frac{-12t}{\rho(s,t)^2} & \frac{60t}{\rho(s,t)^2} \\ \frac{-4(-1+u^2-v^2)}{\rho(u,v)^2} & \frac{-8uv}{\rho(u,v)^2} & \frac{-8u}{\rho(u,v)^2} & \frac{-8u}{\rho(u,v)^2} \\ \frac{-8uv}{\rho(u,v)^2} & \frac{4(1+u^2-v^2)}{\rho(u,v)^2} & \frac{-8v}{\rho(u,v)^2} & \frac{-8v}{\rho(u,v)^2} \end{bmatrix}$$

- From $G(u, v, s, t) = \det(|\mathbf{A}, \mathbf{R}|) = 13{,}824(\mathrm{ut} - \mathrm{sv}) = 0$, we have $t = \frac{s\,v}{u}$
- Substituting t by $\frac{s\,v}{u}$ in Eq. 4 we obtain:

$$F_0(u, v, s) = (11\,s + u)\left(7\,su^2 + 7\,sv^2 - 5\,u\right) = 0,$$

from which we get

$$s_1 = \frac{-1}{11}u, \quad s_2 = \frac{5}{7}\frac{u}{u^2+v^2}.$$

- Finally substituting successively t by $\frac{s\,v}{u}$ and s by s_1 and s_2 in $\mathbf{B}(u, v, s, t)$ we have the parametrization of the bisector

$$\mathbf{B}_1(u, v) = \begin{bmatrix} \frac{11u}{-11+u^2+v^2} \\ \frac{11v}{-11+u^2+v^2} \\ -\frac{9u^2+9v^2+11}{2(-11+u^2+v^2)} \end{bmatrix}, \quad \mathbf{B}_2(u, v) = \begin{bmatrix} \frac{35u}{-5+7u^2+7v^2} \\ \frac{35v}{-5+7u^2+7v^2} \\ -\frac{21u^2+21v^2-25}{2(-5+7u^2+7v^2)} \end{bmatrix}.$$

The untrimmed bisector is composed by two paraboloids with two sheets, see Fig. 2.

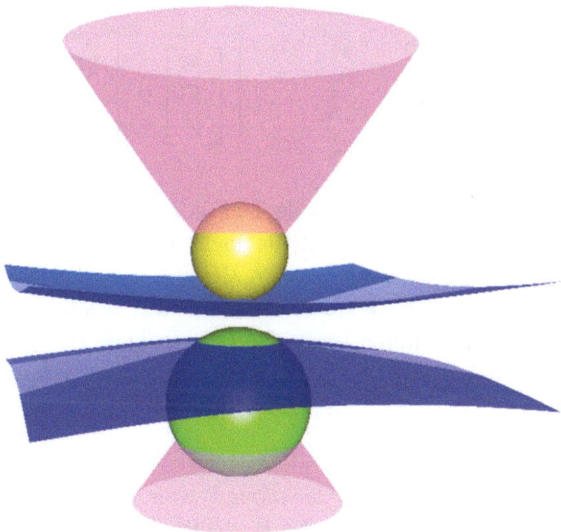

Fig. 2 Bisector of two *spheres*. The *pink* (*light grey*) and the *blue* (*dark gray*) parts correspond to the components $\mathbf{B}_1$ and $\mathbf{B}_2$, respectively

In general, the rationality of the bisector parameterizations $\mathbf{B}_i(u, v)$ depend on the rationality of the norm of the normal vectors of the two surfaces, $\sigma_1(s, t) = \|\partial_s S_1 \times \partial_t S_1\|$ and $\sigma_2(u, v) = \|\partial_u S_2 \times \partial_v S_2\|$. Since σ_1 and σ_2 could be rational functions with non-rational coefficients, i.e.

$$\exists\, \delta_1, \delta_2 \notin \mathbb{Q}, \sigma_1 \in \mathbb{Q}[\delta_1](s, t), \sigma_2 \in \mathbb{Q}[\delta_2](u, v),$$

then the bisector parameterizations $\mathbf{B}_i(u, v)$ could also be rational with non rational coefficients. More precisely if $\sigma_1 \in \mathbb{Q}[\delta_1](s, t)$ and $\sigma_2 \in \mathbb{Q}[\delta_2](u, v)$, then $\mathbf{B}_i \in \mathbb{Q}[\delta_1 \cdot \delta_2](u, v)$.

3.1 Plane-Quadric and Plane-Torus Bisector

Consider a plane parametrized by

$$\mathbf{P}(s, t) = \begin{bmatrix} \alpha_1 & \beta_1 & \gamma_1 \\ \alpha_2 & \beta_2 & \gamma_2 \\ \alpha_3 & \beta_3 & \gamma_3 \end{bmatrix} \begin{bmatrix} s \\ t \\ 1 \end{bmatrix}, \tag{6}$$

where the coefficients α_i, β_i and γ_i are constant, and $\begin{pmatrix} \alpha_1 \\ \alpha_2 \\ \alpha_3 \end{pmatrix} \times \begin{pmatrix} \beta_1 \\ \beta_2 \\ \beta_3 \end{pmatrix} \neq 0$.

Let $\mathbf{S}(u, v) = [S_x(u, v), S_y(u, v), S_z(u, v)]^T$ be the parametrization of a quadric or a torus. For these two surfaces, the Eq. 3 takes the form

$$\begin{bmatrix} \alpha_1 & \alpha_2 & \alpha_3 \\ \beta_1 & \beta_2 & \beta_3 \\ \partial_u S_x & \partial_u S_y & \partial_u S_z \\ \partial_v S_x & \partial_v S_y & \partial_v S_z \end{bmatrix} \begin{bmatrix} X \\ Y \\ Z \end{bmatrix} = \begin{bmatrix} \sum_{i=1}^{3} \alpha_i(\alpha_i s + \beta_i t + \gamma_i) \\ \sum_{i=1}^{3} \beta_i(\alpha_i s + \beta_i t + \gamma_i) \\ \langle S, \partial_u S \rangle \\ \langle S, \partial_v S \rangle \end{bmatrix}. \tag{7}$$

Note that $\mathbf{A}$ does not depend on t, nor in s, while $\mathbf{R}$ is linear in s and t. Thus $G(u, v, s, t) = \det(|\mathbf{A}, \mathbf{R}|)$ is linear in s and t.

Since $\mathbf{P}$ is linear in s and t, $\langle \mathbf{B}(u, v, s, t), 2(\mathbf{S}(u, v) - \mathbf{P}(s, t)) \rangle$ is linear in s and t, and $\|\mathbf{P}(s, t)\|^2$ is quadratic in s and t, whence F (from Eq. 4) is quadratic in s and t. The solution for s of $G(u, v, s, t) = 0$, $s = s(u, v, t)$, will be linear in t. Substituting s by $s(u, v, t)$ in $F(u, v, s, t)$, the resulting expression $F_0(u, v, t) = F(u, v, s(u, v, t), t)$ is quadratic in t. This proves the first part of the following theorem.

Theorem 1 *Consider the plane* $\mathbf{P}$ *with the parametrization* (6)*, and a quadric or a torus* $\mathbf{S}(u, v)$*. Let* $\sigma_p = \|\partial_s \mathbf{P} \times \partial_t \mathbf{P}\|$ *and* $\sigma_s(u, v) = \|\partial_u \mathbf{S} \times \partial_v \mathbf{S}\|$.

1. *The resulting equation:* $F_0(u, v, t) = 0$ *is quadratic in t, whence two components of the parameterization for the bisector are obtained, that might contain a square root of a non-negative expression (originating from* σ_s*).*
2. *If* $\sigma_s(u, v)$ *is rational, then the bisector parameterizations* $\mathbf{B}_i(u, v)$, $i = 1, 2$, *are rational. Furthermore, if* $\sigma_p \notin \mathbb{Q}$ *and* $\sigma_s \in \mathbb{Q}[\delta](u, v)$, *then* $\mathbf{B}_i \in \mathbb{Q}[\boldsymbol{œ}_\mathbf{p} \cdot](u, v)$ *(see Example* 2*).*

Proof We have seen that

$$F_0(u, v, t) = F(u, v, s(u, v, t), t) = \omega_0 t^2 + \omega_1 t + \omega_2 = 0, \tag{8}$$

where ω_0, ω_1, ω_2 are very long expressions depending on α_i, β_i, γ_i, $\partial_u S$, $\partial_v S$, and S.

The discriminant of the Eq. 8 is given by

$$\begin{aligned}\Delta = {} & 4[(\alpha_1\beta_3 - \beta_1\alpha_3)^2 + (-\beta_1\alpha_2 + \alpha_1\beta_2)^2 + (\alpha_2\beta_3 - \alpha_3\beta_2)^2] \\ & [(\partial_u S_y \partial_v S_z - \partial_u S_z \partial_v S_y)^2 + (\partial_u S_z \partial_v S_x - \partial_u S_x \partial_v S_z)^2 \\ & + (\partial_u S_x \partial_v S_y - \partial_u S_y \partial_v S_x)^2][-(\beta_3\beta_2\alpha_3 - \alpha_2\beta_3^2 + \beta_2\alpha_1\beta_1 - \alpha_2\beta_1^2) \\ & (\partial_u S_x \partial_v S_z - \partial_u S_z \partial_v S_x) + (\alpha_2\beta_2\beta_3 - \alpha_3\beta_1^2 + \beta_3\alpha_1\beta_1 - \alpha_3\beta_2^2) \\ & (\partial_u S_x \partial_v S_y - \partial_u S_y \partial_v S_x) - (-\beta_3\beta_1\alpha_3 + \beta_3^2\alpha_1 - \alpha_2\beta_1\beta_2 + \alpha_1\beta_2^2) \\ & (\partial_u S_y \partial_v S_z - \partial_u S_z \partial_v S_y)]^2[(\alpha_2\beta_3 - \alpha_3\beta_2)(S_x - \gamma_1) \\ & - (\alpha_1\beta_3 - \beta_1\alpha_3)(S_y - \gamma_2) + (-\beta_1\alpha_2 + \alpha_1\beta_2)(S_z - \gamma_3)]^2.\end{aligned}$$

After heavy computation and simplifications, we get

$$\begin{aligned}\Delta &= 4\,[\langle \Gamma, (\partial_s P \times \partial_t P)\rangle]^2\,[\langle (\partial_t P \times (\partial_s P \times \partial_t P)), (\partial_u S \times \partial_v S)\rangle]^2 \\ &\quad \|(\partial_s P \times \partial_t P)\|^2\,\|(\partial_u S \times \partial_v S)\|^2 \\ &= 4\,[\langle \Gamma, (\partial_s P \times \partial_t P)\rangle \langle (\partial_t P \times (\partial_s P \times \partial_t P)), (\partial_u S \times \partial_v S)\rangle]^2\,\sigma_p^2 \sigma_s^2(u,v) \geqslant 0,\end{aligned}$$

where $\Gamma = \begin{bmatrix} S_x - \gamma_1 \\ S_y - \gamma_2 \\ S_z - \gamma_3 \end{bmatrix}$. Thus

$$\sqrt{\Delta} = 2\,|\langle \Gamma, (\partial_s P \times \partial_t P)\rangle\,\langle (\partial_t P \times (\partial_s P \times \partial_t P)), (\partial_u S \times \partial_v S)\rangle|\,\sigma_p\,\sigma_s(u,v).$$

Then, the Eq. 8 has two roots:

$$t_i = \frac{-\omega_1 + (-1)^i.\sqrt{\Delta}}{2\omega_0}, \quad i = 1, 2,$$

and substituting s by $s(u,v,t)$, and then t by t_1 and t_2 in $\mathbf{B}(u,v,s,t)$ we obtain the two components of the bisector parametrization

$$\mathbf{B}_1(u,v) = \mathbf{B}(u,v,s(u,v,t_1),t_1) \;\text{ and }\; \mathbf{B}_2(u,v) = \mathbf{B}(u,v,s(u,v,t_2),t_2).$$

Since the quantity

$$|\langle \Gamma, (\partial_s P \times \partial_t P)\rangle\,\langle (\partial_t P \times (\partial_s P \times \partial_t P)), (\partial_u S \times \partial_v S)\rangle|$$

is a rational function, and the norm σ_p of the normal to the plane is a real constant, then the bisector parametrization is rational if the normal vector $\sigma_s(u,v)$ is rational. Furthermore, since in general σ_p is not a rational number and $\sigma_s(u,v)$ could be a rational function with coefficients in an extension field $\mathbb{Q}[\delta]$ (i.e. $\sigma_s(u,v) \in \mathbb{Q}[\delta](u,v)$), the bisector parametrization would be rational with coefficients in the extension field $\mathbb{Q}[\sigma_p.\delta]$. That proves the theorem.

Corollary 1 *The bisector of a plane and a PN-surface is rational. In particular the bisectors of a plane and all quadratic surfaces, except non-circular-cylinders and non-circular-cones, and the bisector of a plane and a torus, are rational.*

Example 2 Consider an oblique plane and a cone, respectively parametrized by

$$P(s,t) = \begin{bmatrix} s \\ t \\ \frac{1}{3}(t+s) \end{bmatrix} \quad\text{and}\quad S(u,v) = \begin{bmatrix} \dfrac{2uv}{1+u^2} \\ \dfrac{v\,(1-u^2)}{u^2+1} \\ v \end{bmatrix}.$$

The norms of the normal vectors are

$$\sigma_p = \frac{\sqrt{11}}{3} \notin \mathbb{Q} \text{ and } \sigma_s = \frac{2\sqrt{2}v}{u^2+1} \in \mathbb{Q}[\sqrt{2}](u, v).$$

Applying the Algorithm 1 we have:

$$G(u, v, s, t) = 13\,su^2 - 2\,tu^2 - 6\,u^2v + 20\,tu + 2\,su - 12\,uv - 4\,t + 6\,v - 7 = 0,$$

$$t = t(u, v, s) = \frac{13\,su^2 - 6\,u^2v + 2\,su - 12\,uv + 6\,v - 7\,s}{2(u^2 - 10\,u + 2)}.$$

Solving $F(u, v, s, t(u, v, s)) = 0$ in terms of the parameter s, we obtain:

$$s_1 = \frac{2\left(-22u - 33u^2 + 44u^3 + 11 + 22u^4 + \sqrt{22}(2 - 12u + 15u^2 - 21u^3 + 2u^4)\right)v}{11(-8u + 3 + 9u^4 - 4u^3 + 12u^2)},$$

$$s_2 = \frac{-2\left(22u + 33u^2 - 44u^3 - 11 - 22u^4 + \sqrt{22}(2 - 12u + 15u^2 - 21u^3 + 2u^4)\right)v}{11(-8u + 3 + 9u^4 - 4u^3 + 12u^2)}.$$

Substituting $t(u, v, s)$, s_1 and s_2 in $B(u, v, s, t)$, we obtain the two components of the bisector parametrization:

$$\mathbf{B_1}(u, v) = \begin{bmatrix} -\dfrac{2uv\left(2\sqrt{22}u^2 - 5\,u^2 - \sqrt{22}u + 6\,u + \sqrt{22} + 1\right)}{3 + 12\,u^2 + 9\,u^4 - 4\,u^3 - 8\,u} \\ \dfrac{v\left(2\sqrt{22}u^4 - 5\,u^4 - \sqrt{22}u^3 + 6\,u^3 - \sqrt{22}u^2 + 6\,u^2 + \sqrt{22}u - 6\,u - \sqrt{22} - 1\right)}{3 + 12\,u^2 + 9\,u^4 - 4\,u^3 - 8\,u} \\ \dfrac{v\left(2\sqrt{22}u^4 + 13\,u^4 - \sqrt{22}u^3 - 2\,u^3 + 3\sqrt{22}u^2 + 20\,u^2 - \sqrt{22}u - 10\,u + 7 + \sqrt{22}\right)}{3 + 12\,u^2 + 9\,u^4 - 4\,u^3 - 8\,u} \end{bmatrix},$$

$$\mathbf{B_2}(u, v) = \begin{bmatrix} \dfrac{2uv\left(2\sqrt{22}u^2 + 5\,u^2 - \sqrt{22}u - 6\,u + \sqrt{22} - 1\right)}{3 + 12\,u^2 + 9\,u^4 - 4\,u^3 - 8\,u} \\ -\dfrac{v\left(2\sqrt{22}u^4 + 5\,u^4 - \sqrt{22}u^3 - 6u^3 - \sqrt{22}u^2 - 6u^2 + \sqrt{22}u + 6u - \sqrt{22} + 1\right)}{3 + 12u^2 + 9u^4 - 4u^3 - 8u} \\ -\dfrac{v\left(2\sqrt{22}u^4 - 13u^4 - \sqrt{22}u^3 + 2\,u^3 + 3\sqrt{22}u^2 - 20u^2 - \sqrt{22}u + 10u - 7 + \sqrt{22}\right)}{3 + 12u^2 + 9u^4 - 4u^3 - 8u} \end{bmatrix}.$$

The two surfaces and the bisector are shown in Fig. 3

3.2 *The Implicit Equation of the Bisector of a Plane and a Rational Surface*

The implicit equation of the bisector of a plane and a low degree rational surface can be easily computed from

- the bisector Eqs. (1–2), or
- the bisector parametrization we computed.

Fig. 3 Bisector of cone and oblique plane

3.2.1 The Implicit Equation from the Bisector Equations

Let $P(s,t)$ be a plane, and let $S(u,v)$ be a surface. The bisector point $B = [x,\ y,\ z]$ satisfies the Eqs. (1–2). Using Cramer's Rule on the two first Equations (1), we obtain:

$$s = s_0 = \frac{\overrightarrow{\Gamma B} \cdot (\partial_t P \times (\partial_s P \times \partial_t P))}{\|\partial_s P \times \partial_t P\|^2},$$

$$t = t_0 = \frac{\overrightarrow{\Gamma B} \cdot (\partial_s P \times (\partial_t P \times \partial_s P))}{\|\partial_s P \times \partial_t P\|^2},$$

where $\overrightarrow{\Gamma B} = \begin{bmatrix} x - \gamma_1 \\ y - \gamma_2 \\ z - \gamma_3 \end{bmatrix}$.

Substituting s and t respectively by s_0 and t_0 in the last three equations in (1–2), we obtain:

$$\begin{cases} F_1(u,v,x,y,z) = \langle B - S(u,v), \partial_u S(u,v)\rangle = 0 \\ F_2(u,v,x,y,z) = \langle B - S(u,v), \partial_v S(u,v)\rangle = 0 \\ F_3(u,v,x,y,z) = \langle B, 2(S(u,v) - P(s_0,t_0))\rangle \\ \qquad\qquad + \|P(s_0,t_0)\|^2 - \|S(u,v)\|^2 = 0 \end{cases}$$

Then, eliminating (using one of the elimination theory tools: Grobner basis, resultants, etc.) the parameters u and v, we obtain the implicit equation of the bisector.

Example 3 Consider the plane and the surface parametrized, respectively, by

$$P(s,t)=\begin{bmatrix} s \\ -4 \\ t \end{bmatrix}, \quad S(u,v)=\begin{bmatrix} \dfrac{1-u^2}{1+u^2} \\ \dfrac{2u}{1+u^2} \\ v \end{bmatrix}.$$

1. $\overrightarrow{\Gamma B}=\begin{bmatrix} x \\ y+4 \\ z \end{bmatrix}$, $s_0=x,\ t_0=z$
2. $F_1(u,v,x,y,z)=-2(1+u^2)(yu^2+2ux-y)$
3. $F_2(u,v,x,y,z)=z-v$
4. $F_3=-(1+u^2)(x^2u^2+v^2u^2-15u^2+z^2u^2-8yu^2+2xu^2-2zvu^2$
   $-4yu-15-2x-8y+x^2+z^2+v^2-2zv)$
5. The implicit equation is given by

$$b(x,y,z)=(-6y-9+x^2)(-10y-25+x^2).$$

3.2.2 The Implicit Equation from the Bisector Parametrization

The bisector parametrization we obtain consists of two components which in general are not rational. If the parametrization is rational, the implicit equation will be computed with the standard method, otherwise we will apply simple techniques to raise the square roots, and then eliminate the parameters u and v.

In the rational case, the implicit equation of the bisector is the product of the two implicit equations of the components.

On the other hand, in the non rational case, as the two components are conjugate (in their expressions), the implicit equation is only obtained from one of the two parametrized components.

Example 4 For the surfaces parametrized by

$$P(s,t)=\begin{bmatrix} s \\ -4 \\ t \end{bmatrix}, \quad \text{and} \quad S(u,v)=\begin{bmatrix} \dfrac{1-u^2}{1+u^2} \\ \dfrac{2u}{1+u^2} \\ v \end{bmatrix},$$

the bisector parametrization is rational, and it is given by

$$B_1(u,v) = \begin{bmatrix} \dfrac{-5(1+u)}{-1+u} \\ \dfrac{10u}{(-1+u)^2} \\ v \end{bmatrix}, \quad B_2(u,v) = \begin{bmatrix} \dfrac{3(-1+u)}{1+u} \\ -\dfrac{6u}{(-1+u)^2} \\ v \end{bmatrix}.$$

1. From $B_1(u, v)$ we obtain:

$$g_1 = x + xu + 3 - 3u, \;\; g_2 = y + 2yu + yu^2 + 6u, \;\; g_3 = z - v,$$

and

$$b_1(x, y, z) = \texttt{eliminate}(\langle g_1, g_2, g_3 \rangle, u, v) = 36y - 6x^2 + 54.$$

2. From B_2 we obtain:

$$b_2(x, y, z) = -10y - 25 + x^2.$$

Then, the implicit equation of the bisector is given by

$$b(x, y, z) = b_1(x, y, z)\, b_2(x, y, z) = (-10y - 25 + x^2)\,(36y - 6x^2 + 54) = 0.$$

3.3 Circular Cylinder-Quadric and Circular Cylinder-Torus Bisectors

In this subsection we apply Algorithm 1 to obtain parameterizations of the bisectors of a circular cylinder **C** of radius r and a surface **S** which could be either a non developable quadric or a torus. Let

$$\mathbf{C}(s,t) = \begin{bmatrix} \dfrac{2rs}{1+s^2} \\ \dfrac{r(1-s^2)}{1+s^2} \\ t \end{bmatrix} \quad \text{and} \quad \mathbf{S}(u,v) = \begin{bmatrix} S_x(u,v) \\ S_y(u,v) \\ S_z(u,v) \end{bmatrix}$$

be the parameterizations of the cylinder and the quadric or torus, respectively.

Applying the algorithm to $\mathbf{C}(s, t)$ and $\mathbf{S}(u, v)$, the equation

$$G(u, v, s, t) = \det(|\mathbf{A}, \mathbf{R}|) = 0$$

is linear in the parameter t and quadratic in the parameter s. More precisely,

$$G(u, v, s, t) = \Theta_0(t)\,(s^2 - 1) + \Theta_1(t)\, s = 0 \tag{9}$$

where

$$
\begin{aligned}
\Theta_0(t) &= -2\,r(\partial_s S_y\,\partial_t S_z - \partial_s S_z\,\partial_t S_y)t - 2\,r(\partial_t S_y\,S_z\,\partial_s S_z \\
&\quad - \partial_s S_y\,S_x\,\partial_t S_x - \partial_s S_y\,S_z\,\partial_t S_z + \partial_t S_y\,S_x\,\partial_s S_x),\\
\Theta_1(t) &= -2\,r(-2\,\partial_s S_x\,\partial_t S_z + 2\,\partial_s S_z\,\partial_t S_x)t - 2\,r(2\,\partial_s S_x\,S_y\,\partial_t S_y \\
&\quad - 2\,\partial_t S_x\,S_y\,\partial_s S_y + 2\,\partial_s S_x\,S_z\,\partial_t S_z - 2\,\partial_t S_x\,S_z\,\partial_s S_z).
\end{aligned}
$$

The method is different for each of the following two configurations:

1. The surface **S** is of revolution, sharing the same axis with the cylinder **C**.
2. The surface **S** is not of revolution or it is not sharing the same axis with the cylinder **C**.

1. S is a surface of revolution, sharing the same axis with the cylinder C: Parametrizing the surface of revolution as follows

$$
\mathbf{S}(u, v) = \begin{bmatrix} A(u)\dfrac{1-v^2}{1+v^2} \\ A(u)\dfrac{2v}{1+v^2} \\ B(u) \end{bmatrix},
$$

the parameter s can be expressed from $G(u, v, s, t) = 0$, in the forms

$$
s_1 = \frac{1-v}{1+v} \quad \text{and} \quad s_2 = -\frac{1+v}{1-v}.
$$

Substituting s_1 and s_2 in $F(u, v, s, t) = 0$ we obtain the following result:

Theorem 2 *Consider the quadric of revolution in the previous paragraph, and let $\sigma_s(u, v) = \|\partial_u S \times \partial_v S\|$. Then*

1. *The resulting equations $F_1(u, v, t) = F(u, v, s_1, t) = 0$ and $F_2(u, v, t) = F(u, v, s_2, t) = 0$, are quadratic in t, whence four components of the parameterization for the bisector* **B** *are obtained, that might contain a square root of a non-negative expression (originating from σ_s).*
2. *If σ_s is rational, then the bisector parametrization* $\mathbf{B}(u, v)$ *is rational. Furthermore, if $\sigma_s \in \mathbb{Q}[\delta](u, v)$, then $\mathbf{B} \in \mathbb{Q}[\delta](u, v)$.*

Proof Substituting s_1 and s_2 in the equation $F(u, v, s, t) = 0$ we obtain

$$
\begin{aligned}
F_1(u, v, t) &= F(u, v, s_1, t) \\
&= 64\,(1 + v^2)^4 (r^2 v^4 + (A(u))^2 v^4 + 4\,r^2 v^3 + 6\,r^2 v^2 \\
&\quad + 2\,(A(u))^2 v^2 + 4\,r^2 v + r^2 + (A\,(u))^2)\Bigg(2\,A\,(u)\,B\,(u)\,\frac{d}{du}B\,(u)
\end{aligned}
$$

$$
\begin{aligned}
&- 2\,A(u)\left(\frac{d}{du}B(u)\right)t - 2\,B(u)\left(\frac{d}{du}B(u)\right)r\\
&+ 2\left(\frac{d}{du}B(u)\right)tr + (A(u))^2\frac{d}{du}A(u)\\
&- 2\left(\frac{d}{du}A(u)\right)rA(u) - \left(\frac{d}{du}r(u)\right)(B(u))^2\\
&+ 2\left(\frac{d}{du}A(u)\right)tB(u)\\
&+ r^2\frac{d}{du}A(u) - \left(\frac{d}{du}A(u)\right)t^2\Big)\frac{d}{du}A(u) = 0,
\end{aligned}
$$

$$
\begin{aligned}
F_2(u, v, t) = F(u, v, s_2, t) = {}& 64(1+v^2)^4(r^2v^4 + (A(u))^2v^4 - 4r^2v^3 + 6r^2v^2\\
&+ 2\,(A(u))^2\,v^2 - 4\,r^2v + r^2 + (A(u))^2)\Big(2\,A(u)\,B(u)\,\frac{d}{du}B(u)\\
&- 2A(u)\left(\frac{d}{du}B(u)\right)t + 2\,B(u)\left(\frac{d}{du}B(u)\right)r\\
&- 2\left(\frac{d}{du}B(u)\right)tr + (A(u))^2\frac{d}{du}r(u)\\
&+ 2\left(\frac{d}{du}A(u)\right)rA(u) - \left(\frac{d}{du}r(u)\right)(B(u))^2\\
&+ 2\left(\frac{d}{du}A(u)\right)tB(u) + r^2\frac{d}{du}A(u)\\
&- \left(\frac{d}{du}A(u)\right)t^2\Big)\frac{d}{du}A(u) = 0.
\end{aligned}
$$

Their respective discriminants with respect to v are given by:

$$
\begin{aligned}
\Delta_1 &= 4\,(A(u) - r)^2\left(\left(\frac{d}{du}B(u)\right)^2 + \left(\frac{d}{du}A(u)\right)^2\right)\\
&= 4\left(1 - \frac{r}{A(u)}\right)^2\sigma_s^2(u, v) \geqslant 0,\\
\Delta_2 &= 4\,(A(u) + r)^2\left(\left(\frac{d}{du}B(u)\right)^2 + \left(\frac{d}{du}A(u)\right)^2\right)\\
&= 4\left(1 + \frac{r}{A(u)}\right)^2\sigma_s^2(u, v) \geqslant 0,
\end{aligned}
$$

and

$$\sqrt{\Delta_1} = 2\left(1 - \frac{r}{A(u)}\right)\sigma_s(u, v), \quad \sqrt{\Delta_2} = 2\left(1 + \frac{r}{A(u)}\right)\sigma_s(u, v),$$

where $\sigma_s(u, v) = \|\partial_u S \times \partial_v S\| = A(u)\sqrt{\left(\frac{d}{du}A(u)\right)^2 + \left(\frac{d}{du}B(u)\right)^2}$. Therefore, for $i = 1, 2$,

$$t_i^1 = \frac{\left(\frac{d}{du}B(u)\right)(A(u) - r) - \left(\frac{d}{du}A(u)\right)B(u) + (-1)^i\left(1 - \frac{r}{A(u)}\right)\sigma_s(u, v)}{\frac{d}{du}A(u)},$$

$$t_i^2 = \frac{\left(\frac{d}{du}B(u)\right)(A(u) + r) - \left(\frac{d}{du}A(u)\right)B(u) + (-1)^i\left(1 + \frac{r}{A(u)}\right)\sigma_s(u, v)}{\frac{d}{du}A(u)}.$$

Substituting successively (s_1, t_i^1) and (s_2, t_i^2) in $\mathbf{B}(u, v, s, t)$, we obtain four components of the bisector parametrization:

$$\mathbf{B}_1(u, v) = \mathbf{B}(u, v, s_1, t_1^1), \quad \mathbf{B}_2(u, v) = \mathbf{B}(u, v, s_1, t_2^1)$$

$$\mathbf{B}_3(u, v) = \mathbf{B}(u, v, s_2, t_1^2), \quad \mathbf{B}_4(u, v) = \mathbf{B}(u, v, s_2, t_2^2)$$

If σ_s is a rational function, then so are t_i^1 and t_i^2, hence the bisector parametrization is rational. That proves the theorem.

Since the norm of every normal vector of all quadrics of revolution and tori is rational, we have:

Corollary 2 *The bisector of a circular-cylinder and a quadric of revolution or a torus sharing the same axis is rational.*

Example 5 Let

$$\mathbf{C}(s, t) = \begin{bmatrix} \dfrac{10\,s}{1 + s^2} \\ \dfrac{5\,(1 - s^2)}{1 + s^2} \\ t \end{bmatrix} \quad \text{and} \quad \mathbf{S}(u, v) = \begin{bmatrix} \dfrac{u\left(-1 + v^2\right)}{\left(-1 + u^2\right)\left(1 + v^2\right)} \\ \dfrac{-2\,uv}{\left(-1 + u^2\right)\left(1 + v^2\right)} \\ \dfrac{-u^2 + u^4 + 1}{(-1 + u)^2\,(u + 1)^2} \end{bmatrix}$$

be a circular cylinder and an elliptic paraboloid. Their bisector parameterizations are given by:

Fig. 4 One component of the bisector of a *cylinder* and a *paraboloid*

$$\mathbf{B}_1(u,v)=\begin{bmatrix}\frac{(11u^2-9)u(-1+v^2)}{(-1+u^2)(1+v^2)(1-u)^2}\\ \frac{-2(11u^2-9)uv}{(-1+u^2)(1+v^2)(1-u)^2}\\ -\frac{4u^4+11u^3+3u^2-9u-6}{(-1+u)^2(1+u)^2}\end{bmatrix},\ \mathbf{B}_2(u,v)=\begin{bmatrix}-\frac{(9u^2-11)u(-1+v^2)}{(-1+u^2)(1+v^2)(1+u)^2}\\ \frac{2(9u^2-11)uv}{(-1+u^2)(1+v^2)(1+u)^2}\\ \frac{6u^4-9u^3-3u^2+11u-4}{(-1+u)^2(1+u)^2}\end{bmatrix},$$

$$\mathbf{B}_3(u,v)=\begin{bmatrix}\frac{(11u^2-9)u(-1+v^2)}{(-1+u^2)(1+v^2)(1+u)^2}\\ \frac{-2(11u^2-9)uv}{(-1+u^2)(1+v^2)(1+u)^2}\\ -\frac{4u^4-11u^3+3u^2+9u-6}{(-1+u)^2(1+u)^2}\end{bmatrix},\ \mathbf{B}_4(u,v)\begin{bmatrix}-\frac{(9u^2-11)u(-1+v^2)}{(-1+u^2)(1+v^2)(1-u)^2}\\ \frac{-2(9u^2-11)uv}{(-1+u^2)(1+v^2)(1-u)^2}\\ \frac{6u^4+9u^3-3u^2-11u-4}{(-1+u)^2(1+u)^2}\end{bmatrix}.$$

See Fig. 4 for a representation of one of the bisector components.

2. The surface S is not of revolution or it is not sharing the same axis with the cylinder C: After expressing t from (9) as $t = t(u, v, s)$, and substituting in $F(u, v, s, t) = 0$ we obtain the following result:

Theorem 3 *Let $S(u, v)$ be a quadric or a torus, $\sigma_s(u, v) = \|\partial_u S \times \partial_v S\|$. If the Algorithm 1 is applied to a circular cylinder and the surface S, then the resulting equation $F_0(u, v, s) = F(u, v, s, t(u, v, s)) = 0$, is quartic in the variable s, and it can be factorized as:*

$$F_0(u, v, s) = F_1(u, v, \sigma_s, s)\, F_2(u, v, \sigma_s, s) = 0$$

where $F_1(u, v, \sigma_s, s)$ and $F_2(u, v, \sigma_s, s)$ are quadratic in s.

Hence, four components of the parameterization for the bisector **B** *are obtained, that in general contain a square root (see Example 6).*

Proof Since the Eq. 9 is linear in t, we can express $t = \frac{B_0}{A_0}$, where

$$\begin{aligned} B_0 &= 2\,\partial_u S_x\, S_y\, \partial_v S_y\, s + 2\,\partial_u S_x\, S_z\, \partial_v S_z\, s - 2\,\partial_v S_x\, S_y\, \partial_u S_y\, s \\ &\quad - 2\,\partial_v S_x\, S_z\, \partial_u S_z\, s + \partial_u S_y\, S_x\, \partial_v S_x + \partial_u S_y\, S_z\, \partial_v S_z \\ &\quad - \partial_v S_y\, S_x\, \partial_u S_x - \partial_v S_y\, S_z\, \partial_u S_z - \partial_u S_y\, S_x\, \partial_v S_x\, s^2 \\ &\quad - \partial_u S_y\, S_z\, \partial_v S_z\, s^2 + \partial_v S_y\, S_x\, \partial_u S_x\, s^2 + \partial_v S_y\, S_z\, \partial_u S_z\, s^2, \\ A_0 &= 2\,\partial_u S_x \partial_v S_z\, s - 2\,\partial_v S_x \partial_u S_z\, s + \partial_u S_y\, \partial_v S_z - \partial_u S_z\, \partial_v S_y \\ &\quad - \partial_u S_y\, \partial_v S_z\, s^2 + \partial_v S_y\, \partial_u S_z\, s^2. \end{aligned}$$

If we substitute $t = t(u, v, s) = \frac{B_0}{A_0}$ in the equation

$$F(u, v, s, t) = 0,$$

and factorize in the extension $\mathbb{R}[\sigma_s](u, v, s)$ we obtain

$$F_0(u, v, s) = F(u, v, s, t(u, v, s)) = F_1(u, v, \sigma_s, s)\, F_2(u, v, \sigma_s, s) = 0 \qquad (10)$$

where F_1 and F_2 are quadratic in s.

If the normal norm σ_s is rational, the factorization (10) will be in the ring $\mathbb{R}[u, v, s]$.

In general, the discriminants of the quadratic polynomials F_1 and F_2 with respect to s are not positive.

Example 6 Let

$$S_1(s, t) = \begin{bmatrix} \dfrac{2s}{1+s^2} + 4 \\ t \\ \dfrac{1-s^2}{1+s^2} \end{bmatrix} \quad \text{and} \quad S_2(u, v) = \begin{bmatrix} \dfrac{2u}{1+u^2+v^2} \\ \dfrac{v}{1+u^2+v^2} \\ \dfrac{2(1-u^2-v^2)}{1+u^2+v^2} \end{bmatrix}$$

be a circular cylinder and an ellipsoid, respectively.

The four components of the bisector parametrization are given by:

$$B_i(u, v) = \begin{bmatrix} -2\,\dfrac{u\left(4\,(s_i\,(1+u^2+v^2))^2 - 4 + 3\,s_i\,(u^2-1+v^2)\right)}{(1+u^2+v^2)(s_i u^2 - 2\,u s_i{}^2 + 2\,u + s_i v^2 - s_i)} \\ \dfrac{v\left((s_i\,(16\,u^2 - 6\,u + 16\,v^2 + 16))^2 - 1 + 15\,u_i\,(u^2-1+v^2)\right)}{(1+u^2+v^2)(s_i u^2 - 2\,u s_i{}^2 + 2\,u + s_i v^2 - s_i)} \\ \dfrac{\left((u_i\,(u^2-1+v^2))^2 - 1\right)(2\,u^2 + 3\,u + 2 + 2\,v^2)}{(1+u^2+v^2)(s_i u^2 - 2\,u s_i{}^2 + 2\,u + s_i v^2 - s_i)} \end{bmatrix}, \quad i = 1, 2, 3, 4,$$

where

Fig. 5 Bisector of *cylinder* and *ellipsoid*

$$s_1 = \frac{a_1 + b_1\sigma + r_1}{16c}, \quad s_2 = -\frac{a_2 + b_2\sigma + r_1}{16c}, \quad s_3 = \frac{a_3 + b_3\sigma + r_2}{16c},$$

$$s_4 = -\frac{a_4 + b_4\sigma + r_2}{16c}, \quad r_1 = \sqrt{\alpha_1 + \beta_1\,\sigma}, \quad r_2 = \sqrt{\alpha_2 + \beta_2\,\sigma},$$

$$\sigma = \sqrt{u^4 + 14\,u^2 + 2\,u^2v^2 + 62\,v^2 + 1 + v^4}, \quad \text{and } a_i, b_i, c, \alpha_i, \beta_i \in \mathbb{R}[u, v, \sigma]$$

are of very large size.

The good component of the bisector (the set of points at equal distance from S_1 and S_2) is shown in Fig. 5.

3.4 Cylinder–Cylinder, Cylinder-Cone, and Cone–Cone Bisector

In this subsection we consider two particular kinds of developable surfaces, which are the cylindrical and conical surfaces. A cylindrical surface (or simply, cylinder) has a parametrization of the form

$$C(s,t) = \begin{bmatrix} f_1(s) + k_1 t \\ f_2(s) + k_2 t \\ f_3(s) + k_3 t \end{bmatrix},$$

where f_1, f_2, f_3, are rational functions, and k_1, k_2, k_3, are real constants. A conical surface (or simply, cone) has a parametrization of the form

$$\hat{C}(u,v) = \begin{bmatrix} v\,g_1(u) \\ v\,g_2(u) \\ v\,g_3(u) \end{bmatrix}$$

where g_1, g_2, g_3, are rational functions.

We analyze the bisector parametrization of different combinations of cylinders and cones. There are three cases. If we apply the Algorithm 1 to two such surfaces, $\mathbf{S}_1(s,t)$, $\mathbf{S}_2(u,v)$, we get the following:

- If $\mathbf{S}_1$ and $\mathbf{S}_2$ are cylinders, then $G(u,v,s,t) = G_0(u,v,s,t) = 0$.
- If $\mathbf{S}_1$ is a cylinder and $\mathbf{S}_2$ is a cone, then $G(u,v,s,t) = v\,G_0(u,v,s,t) = 0$.
- If $\mathbf{S}_1$ and $\mathbf{S}_2$ are cones, then $G(u,v,s,t) = t\,v\,G_0(u,v,s,t) = 0$.

The function G_0 is linear in t and v. There are no points in the bisector arising from $t = 0$ or $v = 0$. If we express $t = t(u,v,s)$, and then substitute in

$$F(u,v,s,t) = 0,$$

we obtain the following result.

Theorem 4 *Suppose $\mathbf{S}_1$ and $\mathbf{S}_2$ are two surfaces of the types mentioned in the previous paragraph. Let $\sigma_1(s,t) = \|\partial_s\mathbf{S}_1 \times \partial_t\mathbf{S}_1\|$ and $\sigma_2(u,v) = \|\partial_u\mathbf{S}_2 \times \partial_v\mathbf{S}_2\|$. If the Algorithm 1 is applied, then*

1. *The equation $F_0(u,v,s) = F(u,v,s,t(u,v,s))$ is quadratic in v. Thus, two components of the parameterization of the bisector $\mathbf{B}$ are obtained, that might contain a square root of a non-negative expression (originating from σ_1 and σ_2).*
2. *If the norms σ_1 and σ_2 are rational, then the parametrization of the bisector is rational. Furthermore if $\sigma_1 \in \mathbb{Q}[\delta_1](s,t), \sigma_2 \in \mathbb{Q}[\delta_2](u,v)$, then $\mathbf{B} \in \mathbb{Q}[\delta_1 \cdot \delta_2](s,u)$.*

Proof We have:

$$F_0(u,v,s) = F(u,v,s,t(u,v,s)) = \omega_0 v^2 + \omega_1 v + \omega_2 = 0,$$

where ω_i are rational functions depending on $\partial f_i,\ f_i,\ k_i,\ i = 1,2,3$.

Computing the discriminant of $F_0(u,v,s) = 0$, we get:

$$\Delta = \omega_1^2 - 4\omega_0\omega_2 = 4\sigma_1^2\sigma_2^2\widetilde{G}^2, \text{ hence } \sqrt{\Delta} = 2\sigma_1\sigma_2\widetilde{G}$$

where $\widetilde{G}$ is a long rational expression depending on $\partial f_i,\ f_i$, and k_i.

We obtain the solutions

$$v_1 = \frac{-\omega_1 - 2\sigma_1\sigma_2\widetilde{G}}{2\omega_0}, \quad v_2 = \frac{-\omega_1 + 2\sigma_1\sigma_2\widetilde{G}}{2\omega_0}.$$

Substituting $t(u,v,s)$, v_1 and v_2 in $\mathbf{B}(u,v,s,t)$, we obtain two components of the bisector parametrization, namely

$$\mathbf{B}_1(s,u) = B(u, v_1(u,s), s, t(u, v_1(u,s))) \text{ and}$$
$$\mathbf{B}_2(s,u) = B(u, v_2(u,s), s, t(u, v_2(u,s))).$$

Fig. 6 Bisector of *elliptic* and *parabolic cylinder*

If σ_1 and σ_2 are rational, then s_1 and s_2 are rational, and therefore the two components of the bisector parametrization are rational. That proves the theorem.

Example 7 Let $C_1(s,t) = \begin{bmatrix} s^2+2 \\ t \\ s \end{bmatrix}$ and $C_2(u,v) = \begin{bmatrix} \dfrac{u}{u^2+1} \\ \dfrac{1-u^2}{u^2+1} \\ v \end{bmatrix}$ be a parabolic cylinder and an elliptic cylinder, respectively. The norms of the respective normal vectors are

$$\sigma_1(s,t) = \sqrt{1+4\,s^2} \text{ and } \sigma_2(u,v) = \frac{\sqrt{14\,u^2+1+u^4}}{\left(u^2+1\right)^2}.$$

The two components of the bisector parametrization are given by:

$$B_1(u,s) = \begin{bmatrix} \dfrac{u\left(-16u\left(u^2+1\right)\left(1+4s^2\right)\left(s^2+2\right)+14\,u^2+1+u^4+4R\right)}{\left(u^2+8su-1\right)\left(-u^2+8su+1\right)\left(u^2+1\right)} \\ -\dfrac{\left(u^2-1\right)\left(-4u\left(u^2+1\right)\left(1+4s^2\right)\left(s^2+2\right)-2u^2\left(-1+24s^2\right)+u^4+1+R\right)}{\left(u^2+8su-1\right)\left(-u^2+8su+1\right)\left(u^2+1\right)} \\ -\dfrac{s\left(-\left(u^2+1\right)\left(\left(u^4+1\right)(5+2s^2)+9u^2(6-4s^2)\right)+2u\left(14u^2+1+u^4\right)+8uR\right)}{\left(u^2+8su-1\right)\left(-u^2+8su+1\right)\left(u^2+1\right)} \end{bmatrix},$$

$$B_2(u,s) = \begin{bmatrix} -\dfrac{u\left(16\,u\left(u^2+1\right)\left(1+4\,s^2\right)\left(s^2+2\right)+u^4-14\,u^2-1+4\,R\right)}{\left(u^2+8\,su-1\right)\left(-u^2+8\,su+1\right)\left(u^2+1\right)} \\ \dfrac{\left(u^2-1\right)\left(4\,u\left(u^2+1\right)\left(1+4\,s^2\right)\left(s^2+2\right)-2\,u^2+48\,s^2u^2-u^4-1+R\right)}{\left(u^2+8\,su-1\right)\left(-u^2+8\,su+1\right)\left(u^2+1\right)} \\ \dfrac{s\left(u^2+1\right)\left(2\,u^4s^2+5\,u^4+32\,s^2u^2+54\,u^2-32\,s^2+5\right)+8\,uR}{\left(u^2+8\,su-1\right)\left(-u^2+8\,su+1\right)\left(u^2+1\right)} \end{bmatrix},$$

with $R = \left(s^2u^2+s^2+2\,u^2+2-u\right)\sqrt{\left(14\,u^2+1+u^4\right)\left(1+4\,s^2\right)}$. See Fig. 6.

Remark 1 This approach can be applied in the same way to deal with the case of two cylinders or two cones.

4 Conclusion

In most cases involving low degree surfaces, the method proposed produces, efficiently, an exact parametrization of the bisector. This includes: plane-quadric, plane-torus, circular cylinder-quadric, circular cylinder-torus, cylinder–cylinder, cylinder-cone, and cone–cone bisectors.

Even if the parameterization obtained is not rational in general, we get a representation that can be used easily for approximation purposes. In addition, the implicit equation of the bisector can be computed from the obtained parametrization.

Acknowledgments The authors are partially supported by the Spanish "Ministerio de Economia y Competitividad" and by the European Regional Development Fund (ERDF), under the Project MTM2011-25816-C02-02, and by the SAGA network.

References

1. Decell, H.P.: An application of the cayley-hamilton theorem to generalized matrix inversion. SIAM Rev. **7**, 526–528 (1965)
2. Diaz-Toca, G.M., Gonzalez-Vega, L., Lombardi, H.: Generalizing Cramer's rule: solving uniformly linear systems of equations. SIAM J. Matrix Anal. Appl. **27**(3), 621–637 (2005)
3. Dutta, D., Hoffman, C.: On the skeleton of simple csg objects. ASME J. Mech. Des. **115**, 87–94 (1993)
4. Elber, G., Kim, M.-S.: Computing rational bisectors. IEEE Comput. Graphics Appl. **19**(6), 76–81 (1999)
5. Elber, G., Kim, M.-S.: Rational bisectors of CSG primitives. In: Proceedings of 5th ACM/IEEE Symposium on Solid Modeling and Applications, pp. 246–257. Ann Arbor, Michigan, June 1999
6. Elber, G., Kim, M.-S.: A computational model for non-rational bisector surfaces: curve-surface and surface-surface bisectors. In: Proceedings of Geometric Modeling and Processing 2000, pp. 364–372. Hong Kong, Apr 2000
7. Farouki, R.T., Johnstone, J.K.: The bisector of a point and a plane parametric curve. Comput. Aided Geom. Des. **11**(2), 117–151 (1994)
8. Farouki, R.T., Johnstone, J.K.: Computing point/curve and curve/curve bisectors. In: Fisher, R.B. (ed.) Design and Application of Curves and Surfaces (Mathematics of Surfaces V), Oxford University Press, Oxford, pp. 327–354 (1994)
9. Kim, M-S., Elber, G., Seong, J.-K.: Geometric computations in parametric space.In: Spring Conference on Computer Graphics, pp. 12–14. Bundmerice Castle, Slovak Republic (2005)
10. Peternell, M.: Geometric properties of bisector surfaces. Graph. Models **62**(3), 202–236 (2000)
11. Peternell, M.: Sphere-geometric aspects of bisector surfaces. In: Proceedings of AGGM 2006, pp. 107–112. Barcelona (2006)

Computer Aided Algebraic Geometry: Constructing Surfaces of Genus Zero

Roberto Pignatelli

Abstract Everybody knows that mathematics has a key role in the development of the modern technology. It is less known that the modern technology gives something back to mathematics. In this note we give an account on how the combination of classical results as the Riemann Existence Theorem with the use of computers and computational algebra programs answered interesting old-standing problems in classical algebraic geometry, namely regarding the construction and the classification of new surfaces of general type. We also give a full list of the surfaces constructed with this method up to now, and present the next challenges on the subject.

1 Introduction

Most people divide the mathematicians in two subsets, pure mathematicians and applied mathematicians. They imagine the mathematicians in the first class as old, thinking about very abstract problems, and working only with pen and paper, whence the mathematicians in the second class are expected to be younger, interested only in very concrete questions and spending most of their time in front of the screen of a computer or some other electronic devices.

I know indeed many mathematicians who fit perfectly in one of the two models above. But one can do interesting research without corresponding to any of them. In particular, one can study classical problems, problems considered already many decades ago, sometimes more than a century ago, and use the new technologies to answer those questions.

I am mainly interested in the study of smooth compact complex algebraic surfaces, which I will for short denote just by surfaces from now on. This theory essentially started in the XIX century, thanks to the impulse of the *Italian school* leaded by the

R. Pignatelli (✉)
Department of Mathematics, University of Trento, 38127 Trento, Italy
e-mail: Roberto.Pignatelli@unitn.it

R. De Amicis and G. Conti (eds.), *Future Vision and Trends on Shapes, Geometry and Algebra*, Springer Proceedings in Mathematics & Statistics 84,
DOI: 10.1007/978-1-4471-6461-6_6,

genius of G. Castelnuovo and F. Enriques. One of the most interesting questions at that time was the one posed by Max Noether, which we can express in modern terms as *Is a surface of genus zero birational to a projective plane*? The (geometric) genus of a surface is the dimension of the vector space of the holomorphic 2-forms. Max Noether knew that the analogous question in dimension one has positive answer, but could not settle the two-dimensional case.

Enriques answered negatively to this question few years later before the end of the XIXth century, by his beautiful construction of a sextic surface double along the edges of a tetrahedron [15].

Many similar questions have been answered in the next century. *Is a surfaces homeomorphic to a projective plane isomorphic to it?* Yes [24]. *Is a surface homeomorphic to the product of two projective lines isomorphic to it?* No [18]. *Is a surface with the same homology groups of the projective plane isomorphic to it?* No [20]. *Can we list all surfaces with the same homology groups of the projective plane?* Yes [11, 22].

The answer to the last question is a clear example of what I mentioned at the beginning. By a theorem of Yau [25] each surface with the same Betti numbers of the projective plane is rigid and a quotient of the open ball by an automorphism group acting properly discontinuously. It follows that there are only finitely many surfaces with those Betti numbers, but to get a full list of them a computer was necessary, together with four clever mathematicians, and now we can say

Theorem [11] *There are exactly* 101 *pairwise not isomorphic surfaces with the Betti numbers of the projective planes.*

All these surfaces have genus zero. More precisely the 100 surfaces different from the projective planes are surfaces of general type, which means that the canonical bundle, the bundle of the two forms, is as *positive* as it can be: tensoring it with itself k times, and leaving k go to infinity, one gets bundles with many holomorphic global sections, and more precisely forming a vector space whose dimension grows as the square of k. Despite this *positivity*, the genus (which is the dimension for $k = 1$) is as small as possible, zero!

The surfaces of general type of genus zero enter indeed in all the questions above, and in most of the open analogous questions. Hirzebruch [18] produced countably many surfaces not pairwise isomorphic which are homeomorphic to the product of two lines, now known as *Hirzebruch surfaces*. The natural question follows: *Is every surface homeomorphich to the product of two projective lines isomorphic to one of the Hirzebruch surfaces*? It is known that if there is a surface homeomorphic to the product of two lines not in Hirzebruch's list, then it is a surface of general type and genus zero.

There are many more motivations to study this class of surfaces coming from different subjects as surface theory, differential geometry, and cycle theory: see the survey [5]. We just add that recently also their derived category has shown to be very special, and have been used to show the existence of certain special objects, the phantoms, which were conjectured not to exist, see [10].

A full classification of these surfaces is still out of reach. Bauer et al. [5] describes the state of the art of this research one year ago. Some more constructions appeared last year: [1, 2, 8, 14, 21, 23].

A strategy for constructing these surfaces origins from an idea of Beauville [7]. Beauville considers the quotient of a product of two curves by a finite group G of automorphisms acting freely. These surfaces are now known as *surfaces isogenous to a product*. If the order of G is $(g_1-1)(g_2-1)$ then the genus p_g equals the irregularity q, which is the dimension of the vector space of the holomorphic 1-forms, which is easy to compute and vanishes in most cases. This gives minimal surfaces with ample canonical class K of self-intersection 8.

The self-intersection of the canonical class of a minimal surface of general type is an integer which may vary among 1 and 9, and is a topological invariant, so the surfaces isogenous to a product can touch only few topological types of surfaces of genus zero, and it is natural to ask if we can generalize the construction to be able to touch also other values of K^2.

The generalization we suggest is to consider quasi-étale quotients, which are quotients of a product of two curves by a finite group G acting freely out of finitely many points. The quotient is singular and we need to consider a resolution of its singularities. This gives many new examples of surfaces of genus zero, as we will see in the following.

In the next sections we give an account on the results about surfaces of genus zero which have been found up to now with this method. In the following section we explain the details of the method, and in Sect. 4 the algorithm we implemented in MAGMA to find all these surfaces. Finally, in the last section we list the most interesting open problem on the subject.

2 Quasi-Étale Surfaces

We start by giving the main definitions.

Definition A surface X is *isogenous to a product* if X is the quotient of a product $C_1 \times C_2$ of two algebraic curves by the action of a finite group of automorphisms G acting freely.

We are mainly interested in the following generalization.

Definition A surface X is a *a quasi-étale quotient* if X is the quotient of the product $C_1 \times C_2$ by the action of a finite group of automorphisms G acting freely out of a finite subset of $C_1 \times C_2$.

We use the word quasi-étale because the natural projection $\pi\colon C_1 \times C_2 \to X$ is quasi-étale in the sense of [13]. A quasi-étale quotient, if not isogenous to a product, is never smooth, but it is singular at finitely many points, the images of the *small* orbits of the G-action on $C_1 \times C_2$, the orbits of cardinality smaller than the order of

G. Since we are interested in the classification of smooth surfaces of general type, we introduce the following

Definition A surface S is a *a quasi-étale surface* if it is the minimal resolution of the singularities of a quasi-étale quotient of X.

Catanese [12] has noticed that

1. For every quasi-étale quotient X there are infinitely many triples (C_1,C_2,G) such that $X = (C_1 \times C_2)/G$, but the triple producing X with G of minimal order is unique up to isomorphism, and we call it *a minimal realization* of X; in this case we say that the action of G on $C_1 \times C_2$ is *minimal.*
2. A minimal finite group action on $C_1 \times C_2$ is either *unmixed*, which means that it is the action *diagonally* induced by two faithful actions of G respectively on C_1 and C_2 or *mixed*, in which case $C := C_1 = C_2$, half of the elements of G exchange the two factors, and the other half form a subgroup $G^0$of index 2 acting minimally unmixed on $C \times C$.

Working on Catanese's remarks, Frapporti proved that the quasi-étale assumption is algebraically very simple.

Theorem [16] *Assume that G acts minimally on $C_1 \times C_2$.*
Then $X = (C_1 \times C_2)/G$ is not a quasi-étale quotient (equivalently, the natural map $\pi\colon C_1\times C_2 \to X$ is not quasi-étale) if and only if the action is mixed and the exact sequence

$$1 \to G^0 \to G \to Z_{/2Z} \to 1$$

does not split (equivalently there are no elements of G-G^0 of order 2).

This allows to check very simply if an action is quasi-étale. Having an explicit quasi-étale action, the singular points of X are finitely many. A careful analysis of them allows computing all the invariants of the quasi-étale surfaces S.

We are mainly interested in the opposite argument: fixed the invariants, can we produce surfaces with those invariants? And, more ambitiously, can we give a full list of all quasi-étale surfaces with those invariants?

The answer to both questions are positive, but the latter only in a weak sense. Indeed we have written a computer program in the MAGMA [9] language producing in principle (assuming to have enough time and RAM) all quasi-étale surfaces of general type with given genus, irregularity and K^2. Still, this is not fully satisfactory for us. We wish to have a full list of surfaces of general type of genus zero, and these have automatically irregularity zero, but there is in principle no lower bound for K^2.

As mentioned in the introduction the self-intersection of the canonical class of the minimal model of a surface of general type of genus zero is a number among 1 and 9, but since blowing up a point the selfintersection of the canonical class drops by one, the K^2 of a surfaces of general type with genus zero is not bounded from below.

Anyway, quasi-étale surfaces are not far from being minimal. Indeed, it is easy to see that the canonical class of a surface of general type isogenous to a product is ample, and therefore the surface is minimal, so in this case we just need to run the program for all values of K^2 from 1 to 9. The result can be shortly summarized in the following.

Theorem [3, 16] *There are exactly* 18 *irreducible families of surfaces of general type of genus zero isogenous to a product, forming* 18 *irreducible components of the moduli space of the surfaces of general type.*

Unfortunately this is not true for quasi-étale surfaces: Mistretta and Polizzi [19] found the first example of a not minimal quasi-étale surface. Running our programs for K^2 from 1 to 9 we proved

Theorem [4, 6, 17] *There are exactly* 88 *irreducible families of minimal quasi-étale surfaces,*

and

Theorem [6, 17] *There is exactly one quasi-étale surface S of general type with positive K^2which is not minimal; it is a surface with $K^2 = 1$ whose minimal model has $K^2 = 3$.*

We have organised the 89 families in the Table 1

3 Constructing Curves with a Group Action

By Riemann Existence Theorem, to give an action of a group G^0 onto a curve C is equivalent to give

- The quotient curve C/G^0;
- Finitely many points $p_1, \ldots, p_r$, in C/G^0, the branching points of the projection map $C \to C/G^0$;
- A suitable set of generators of the fundamental group of the complement of these points in C/G^0, including loops $\gamma_1, \ldots, \gamma_r$ such that each γ_i is a small circle around p_i;
- A suitable set of generators of G^0; these are the images of the above generators of the fundamental group of the complement of the p_i by the standard monodromy map.

The signature of this system of generators is the unordered list of the orders of the elements of this last set of generators which are the images of the γ_i by the monodromy map.
So, to construct an unmixed quotient surface, we have to choose two sets of points $p_1, \ldots, p_r$, in C_1/G^0 and $q_1, \ldots, q_r$, in C_2/G^0, corresponding systems of generators as

Table 1 All families of minimal quasi-étale surfaces of general type of genus zero

K^2	m/u	g_1	g_2	o(G)	Id(G)	Sing(X)	Signature(s)	$\lvert\pi_1\rvert$	N
8	U	21	4	60	5	–	$2,5^2;3^4$	∞	1
8	U	16	5	60	5	–	$3^2,5;2^5$	∞	1
8	U	13	6	60	5	–	$2^3,3;5^3$	∞	1
8	U	25	3	48	48	–	$2,4,6;2^6$	∞	1
8	U	9	5	32	27	–	$2^3,4;2^2,4^2$	∞	1
8	U	6	6	25	2	–	$5^3;5^3$	∞	1
8	U	13	3	24	12	–	$3,4^2;2^6$	∞	1
8	U	9	3	16	11	–	$2^3,4;2^6$	∞	1
8	U	5	5	16	3	–	$2^2,4^2;2^2,4^2$	∞	1
8	U	5	5	16	14	–	$2^5;2^5$	∞	1
8	U	4	4	9	2	–	$3^4;3^4$	∞	1
8	U	5	3	8	5	–	$2^6;2^5$	∞	1
8	M	17		256	3678	–	4^3	∞	3
8	M	17		256	3679	–	4^3	∞	1
8	M	9		64	92	–	2^5	∞	1
6	U	3	7	16	11	$2 1/2$	$2^3,4;2^4,4$	∞	1
6	U	19	3	48	48	$2 1/2$	$2^3,4;2^4,4$	∞	1
6	U	19	8	168	42	$2 1/2$	$2^3,4;2^4,4$	84	2
6	U	19	16	360	118	$2 1/2$	$2^3,4;2^4,4$	60	2
6	U	19	11	240	189	$2 1/2$	$2^3,4;2^4,4$	120	1
6	U	4	16	60	5	$2 1/2$	$2^3,4;2^4,4$	∞	1
5	U	3	17	48	48	$1/3 + 2/3$	$2,4,6;2^4,3$	∞	1
5	U	9	9	96	227	$1/3 + 2/3$	$2^3,3;3,4^2$	32	1
5	U	9	3	24	12	$1/3 + 2/3$	$2^4,3;3,4^2$	∞	1
5	U	9	11	120	34	$1/3 + 2/3$	$2,5,6;3,4^2$	40	1
5	U	9	6	60	5	$1/3 + 2/3$	$3,5^2;2^3,3$	40	1
5	U	9	6	60	5	$1/3 + 2/3$	$3,5^2;2^3,3$	20	1
5	U	9	5	48	48	$1/3 + 2/3$	$4^2,6;2^3,3$	∞	1
4	U	13	3	48	48	$4 1/2$	$2^2,4^2;2,4,6$	∞	1
4	U	3	3	8	5	$4 1/2$	$2^5;2^5$	∞	1
4	U	13	3	48	48	$4 1/2$	$2^5;2,4,6$	∞	1
4	U	4	4	18	4	$4 1/2$	$2^2,3^2;2^2,3^2$	27	1
4	U	5	5	32	27	$4 1/2$	$2^3,4;2^3,4$	32	1
4	U	3	3	8	2	$4 1/2$	$2^2,4^2;2^2,4^2$	∞	1
4	U	7	3	24	12	$4 1/2$	$2^5;3,4^2$	∞	1
4	U	4	4	18	3	$4 1/2$	$3,6^2;2^2,3^2$	∞	1
4	U	21	4	120	34	$4 1/2$	$3,6^2;2,4,5$	∞	1
4	U	4	3	16	11	$4 1/2$	$2^5;2^3,4$	∞	1
4	U	4	11	60	5	$4 1/2$	$2,5^2;2^2,3^2$	15	1
4	U	10	25	360	118	$2/5 + 3/5$	$2,4,5;3^2,5$	6	1
4	U	5	25	160	234	$2/5 + 3/5$	$2,4,5;4^2,5$	8	3
4	U	10	5	60	5	$2/5 + 3/5$	$2^3,5;3^2,5$	12	1
4	M	5		32	7	$4 1/2$	2^5	32	1

(continued)

Table 1 (continued)

K^2	m/u	g_1	g_2	o(G)	Id(G)	Sing(X)	Signature(s)	$\|\pi_1\|$	N
4	M	5		32	22	$4 1/2$	2^5	∞	1
4	M	9		128	836	$4 1/2$	4^3	32	1
3	U	10	25	360	118	$1/5 + 4/5$	$2,4,5;3^2,5$	6	1
3	U	5	25	160	234	$1/5 + 4/5$	$2,4,5;4^2,5$	8	3
3	U	10	5	60	5	$1/5 + 4/5$	$2^3,5;3^2,5$	12	1
3	U	11	3	48	48	$2 1/2 + 1/3 + 2/3$	$2^2,3,4;2,4,6$	8	1
3	M	?		64	42	$3/8 + 5/8$	$2^3,8$	8	1
2	U	7	3	48	48	$6 1/2$	$2^3,4;2,4,6$	4	1
2	U	4	4	36	10	$6 1/2$	$2,6^2;2^3,3$	3	1
2	U	4	6	60	5	$6 1/2$	$2,5^2;2^3,3$	5	1
2	U	3	3	16	2	$6 1/2$	$4^3;4^3$	8	1
2	U	3	3	16	11	$6 1/2$	$2^3,4;2^3,4$	8	1
2	U	3	22	168	42	$6 1/2$	$2,3,7;4^3$	4	2
2	U	11	3	120	34	$6 1/2$	$2,6^2;2,4,5$	3	1
2	U	10	7	168	42	$2 1/2 + 1/4 + 3/4$	$2,4,7;3^2,4$	3	2
2	U	10	16	360	118	$2 1/2 + 1/4 + 3/4$	$2,4,5;3^2,4$	3	2
2	U	16	4	120	34	$2 1/2 + 1/4 + 3/4$	$3,4,6;2,4,5$	3	2
2	U	5	3	24	12	$2(1/3 + 2/3)$	$2^2,3^2;3,4^2$	8	1
2	U	3	5	24	13	$2(1/3 + 2/3)$	$2,6^2;2^2,3^2$	8	1
2	U	6	6	75	2	$2(1/3 + 2/3)$	$3^2,5;3^2,5$	5	2
2	U	5	6	60	5	$2(1/3 + 2/3)$	$3^2,5;2^3,3$	4	1
2	M	3		16	3	$6 1/2$	2^5	8	1
2	M	5		64	82	$6 1/2$	4^3	8	1
2	M	5		64	32	$1/2 + 2 3/4$	$2^3,4$	4	1
2	M	7		36	9	$1/2 + 2 3/4$	$2^2,3^2$	3	1
2	M	9		128	1535	$3 1/2 + 2 1/4$	$2^3,4$	8	1
2	M	17		768	1083540	$2(1/3 + 2/3)$	$3^2,4$	4	1
2	M	17		768	1083541	$2(1/3 + 2/3)$	$3^2,4$	4	1
1	U	3	3	24	12	$4 1/2 + 1/3 + 2/3$	$2^3,3;3,4^2$	4	1
1	U	3	15	168	42	$4 1/2 + 1/3 + 2/3$	$2,3,7;3,4^2$	2	1
1	U	3	5	48	48	$4 1/2 + 1/3 + 2/3$	$2, 4,6;2^3,3$	2	1
1	M	3		32	6	$2 1/2 + 2 3/4$	$2^3,4$	4	1

Legend: every rows correspond to an irreducible family of quasi-étale surfaces. In the first column we put K^2 of the surface. In the second column we put U or M according if the action is unmixed or mixed. In the third column we put the genus of the first curve, C in the mixed case, C_1 in the unmixed case. In the fourth column we put the genus of C_1 in the unmixed case, nothing in the mixed case. In the fifth column we put the order of the group G. In the sixth column we put the identifier of G among the groups of its order in the MAGMA database. In the seventh column we put the basket of singularities; since all the singular points are cyclic quotient singularities we represent them as fractions so that, for example, $3 1/2 + 2 1/4$ means that there 5 singular points, 3 quotient singularities of type 1/2 and 2 of type 1/4. In the eighth column we put the signatures (only one in the unmixed case) in exponential notation, so that $2^3, 4$ correspond to $\{2, 2, 2, 4\}$. In the ninth column we put the order of the fundamental group of the resulting surfaces. In the last column we put the number of families up to isomorphisms

above, and that's all. To give a mixed quasi-étale quotient it is enough to give [16] a set of points, a set of generators of both the fundamental group and the group G^0, thus determining the action of G^0 on the first factor of $C \times C$, and then the unsplit extension $1 \to G^0 \to G \to Z_{/2Z} \to 1$.
The formulas for the invariants of the resulting surface are simple to compute. For the irregularity

$$q = g(C_1) + g(C_2)$$

in the unmixed case and

$$q = g(C)$$

in the mixed case.

The formulas for the other invariants are slightly more complicated. For each singular point x in X we can compute its analytic type, which only depends on the stabilizers of the corresponding points in $C_1 \times C_2$. There are two positive rational numbers k_x and b_x [6, 17] depending only on the analytic type of the singularity such that

$$\begin{aligned} K^2 &= 8(g(C_1)-1)(g(C_2)-1)/|G| + \Sigma_x k_x \\ p_g &= K^2/8 - 1 + q + \Sigma_x b_x \end{aligned}$$

We can now construct all quasi-étale surfaces and compute their numerical invariants.

4 Constructing Surfaces

Constructing random surfaces, it is not very likely to give anything interesting. From the discussion in the introduction is clear that we are interested in a systematic procedure for constructing surfaces with prescribed invariants, for example for genus zero.

The formulas in the last section can be "reversed" to an algorithm, which does exactly that, as follows. Note that the unmixed and the mixed case have to be treated separately (and indeed we have two distinct programs) but the idea of the algorithm is the same.

- Choose the invariants p_g, q and K^2 (these integers are the input).
- In the unmixed case, consider all possible pairs of genera of the quotient curves with $q = g(C_1/G) + g(C_2/G)$; in the mixed case $q = g(C/G^0)$.
- Determine all possible "baskets" of singularities x such that $\Sigma_x b_x = K^2/8 - 1 + q - p_g$; it is not difficult to show, from the explicit formula for b_x, that there are only finitely many possible baskets like that, and there is a simple algorithm to compute this list.

- There are some inequalities, which in both cases, once fixed the invariants and the basket of singularities, restrict the possible signatures to a finite set: our algorithm computes, for each basket, the corresponding list.
- Hurwitz formula computes the order of G^0 (which coincides with G in the unmixed case) from the invariants and the signatures; then (for each pairs of signatures in the unmixed case, for each signature in the mixed case) we run a search among all groups of the right order for (pairs of) systems of generators of the prescribed signature(s). Of course this can be done only if the order of the group is small enough; up to now we could exclude all other cases by group theoretical arguments.
- In the unmixed case, for each such system of generators we compute all unsplit extensions of degree 2 of G^0.
- Each of the output we have gives a surface, whose singularities we can compute. If the basket of singularities is exactly the one prescribed by the basket we are considering, then we have a surface with the invariants we have fixed at the beginning. Else we just throw it in the waste.

Both algorithms (in the mixed and unmixed case) are now implemented in MAGMA, and available upon request to the author, although the algorithm for the unmixed case works at the moment only in the regular case ($q = 0$). These programs produce, for $p_g = 0$ and $K^2 > 0$, the 89 families in Sect. 2.

5 Problems

Problem 1: Bound K^2 from below by a function of p_g and q.

An upper bound is provided by Bogomolov-Miyaoka-Yau inequality, while there is no lower bound in general, but we have some evidences that there should be some lower bound for surfaces constructed in this way. We have some conjectures and only partial results in this direction. This would allow us to obtain the full list of all quasi-étale surfaces of genus zero.

Problem 2: If $q = 1$, the Albanese map is a morphism onto an elliptic curve. Compute the genus of general fibre.

This is simple in the unmixed case, where we can always assume $g(C_1) = 1$ and then the Albanese map is an isotrivial fibration with fibre C_2. But it is absolutely non trivial in the mixed case. Once solved this problem, we could use our programs to test some interesting existence conjecture in the irregular case.

Problem 3: Simplify the algorithm!

The program complexity is very big; if we drop K^2 in the input by 1 the time necessary to complete the computation is multiplied approximately by 30. We run the program up to $K^2 = -2$, but with this program we cannot go much further. Moreover every tentative to use the program for $p_g > q$ ended, up to now, with a computer running out of memory.

Acknowledgments The author is member of the FCT project PTDC/MAT/111332/2009 *Moduli spaces in algebraic geometry*, the PRIN 2010–2011 project *Geometria delle varietà algebriche*, and the Futuro in Ricerca 2012 project *Spazi di moduli e applicazioni*. All these groups supported this research.

References

1. Bauer, I., Catanese, F.: Burniat surfaces III: deformations and automorphisms and extended Burniat surfaces. http://arxiv.org/abs/1012.3770 (2010). Accessed 16 Dec 2010
2. Bauer, I., Catanese, F.: Burniat-type surfaces and a new family of surfaces with $p_g = 0$ and $K^2 = 3$. http://arxiv.org/abs/1209.1980 (2012). Accessed 10 Sep 2012
3. Bauer, I., Catanese, F., Grunewald, F.: The classification of surfaces with $p_g = q = 0$ isogenous to a product of curves. Pure Appl. Math. Q. **4**, 547–586 (2008)
4. Bauer, I., Catanese, F., Grunewald, F., Pignatelli, R.: Quotients of products of curves, new surfaces with $p_g = 0$ and their fundamental groups. Am. J. Math. **134**, 993–1049 (2012)
5. Bauer, I., Catanese, F., Pignatelli, R.: Surfaces of general type with geometric genus zero: a survey. Complex and Differential Geometry. Springer Proceedings in Mathematics, vol. 8, pp. 1–48. Springer, Heidelberg (2011)
6. Bauer, I., Pignatelli, R.: The classification of minimal product-quotient surfaces with $p_g = 0$. Math. Comp. **81**, 2389–2418 (2012)
7. Beauville, A.: Complex Algebraic Surfaces. LMS Lecture Note Series. 68 Cambridge University Press, Cambridge (1983).
8. Bini, G., Favale, F.F., Neves, J., Pignatelli, R.: New examples of Calabi-Yau threefolds and genus zero surfaces. Commun. Contemp. Math. **16**(2), 20 (2014) 1350010
9. Bosma, W., Cannon, J., Playoust, C.: The magma algebra system, i. the user language. J. Symbolic. Comput. 24, 235–265 (1997).
10. Boehning, C., Graf von Bothmer, H.C., Sosna, P.: On the derived category of the classical Godeaux surface. http://arxiv.org/abs/1206.1830 (2012). Accessed 8 Jun 2012
11. Cartwright, D.I., Steger, T.: Enumeration of the 50 fake projective planes. C. R. Math. Acad. Sci. Paris **348**, 1113 (2010)
12. Catanese, F.: Fibred surfaces, varieties isogenous to a product and related moduli spaces. Amer. J. Math. **122**, 1–44 (2000)
13. Catanese, F.: Q.e.d. for algebraic varieties with an appendix by Soenke Rollenske. J. Differential Geom. **77**, 43–75 (2007)
14. Chen, Y.: A new family of surfaces with $K^2 = 7$ and $p_g = 0$. http://arxiv.org/abs/1210.4633 (2012). Accessed 17 Oct 2012
15. Enriques, F.: Memorie scelte di Geometria, vol. I. Zanichelli, Bologna (1956)
16. Frapporti, D.: Mixed surfaces, new surfaces of general type with $p_g = 0$ and their fundamental group. http://arxiv.org/abs/1105.1259 (2011). Accessed 4 May 2012
17. Frapporti, D.: Mixed quasi-étale surfaces, new surfaces of general type with $p_g = 0$ and their fundamental group. Ph. D. Thesis, University of Trento (2012).
18. Hirzebruch, F.: Ueber eine Klasse von einfach-zusammenhaengenden komplexen Mannigfaltigkeiten. Math. Ann. **124**, 77–86 (1951)
19. Mistretta, E., Polizzi, F.: Standard isotrivial fibrations with $p_g = q = 1$. II. J. Pure Appl. Algebra **214**, 344–369 (2010)
20. Mumford, D.: An algebraic surface with K ample, $(K^2) = 9$, $p_g = q = 0$. Amer. J. Math. **101**, 233–244 (1979)
21. Neves, J., Pignatelli, J.: Unprojection and deformations of tertiary Burniat surfaces. Ann. SNS. **XIII**(1), 225–254 (2014)
22. Prasad, G., Yeung, S.: Fake projective planes. Invent. Math. **168**, 321–370 (2007)

23. Rito, C.: Involutions on surfaces with $p_g = q = 0$ and $K^2 = 3$. Geom. Dedicata **157**, 319–330 (2012)
24. Yau, S.T.: Calabi's conjecture and some new results in algebraic geometry. Proc. Nat. Acad. Sci. USA **74**, 1798–1799 (1977)
25. Yau, S.T.: On the Ricci curvature of a compact Kaehler manifold and the complex Monge-Ampère equation. J. Comm. Pure Appl. Math. **31**, 339–411 (1978)

Elastic Bone Transformation for Realistic Facial Animation in WebGL

Andoni Mujika, Nagore Barrena, Sara García and David Oyarzun

Abstract This chapter describes the mathematical model that will be used to animate a virtual face in a project called SPEEP, a project that makes use of this virtual face to teach the pronunciation of foreign languages to students. The mathematical model is based on the well-known Skeleton Subspace Deformation, but an elastic layer is inserted in the generation of bone's transformations. Besides, the whole process that will be followed in the project is described, from the definition of the skeleton structure and the training of the parameters of the model to its application in a WebGL environment.

1 Introduction

The utilization of avatars, 3D virtual characters, in films like Toy Story or Shrek is widely widespread and it is the same for computer and console games. That is why the animation of 3D characters is a very mature technology. Specifically, the facial animation of virtual characters is increasingly realistic due to its importance in 3D applications, especially in those where the avatar talks to the user. Indeed, a study made by Hodgins et al. [1] proved that an audiovisual content with virtual characters looses emotional value if the facial animation is not of high quality.

Although historically the intervention of modelers has been needed to obtain realistic results, nowadays their work could be restricted to a final stage, since actual facial motion capture systems and new 3D engines give a considerably realistic animation. The modelers only have to correct the errors of the system. Nevertheless, in real-time applications modelers cannot touch the animation and realism must be obtained directly from the capture system. In the case of Visual Speech Synthesis, where the movements and the voice of the character are synthesized from plain text,

A. Mujika (✉) · N. Barrena · S. García · D. Oyarzun
Vicomtech-ik4, Mikeletegi Pasealekua, 57, 20009 Donostia-San Sebastian, Spain
e-mail: amujika@vicomtech.org

R. De Amicis and G. Conti (eds.), *Future Vision and Trends on Shapes, Geometry and Algebra*, Springer Proceedings in Mathematics & Statistics 84,
DOI: 10.1007/978-1-4471-6461-6_7,

the motion capture and the corrections must be done before the application is run. Then, the animation is generated automatically in real-time.

In SPEEP, the project for foreign language pronunciation learning partly funded by the Basque Government, which we are presenting in this chapter, both problems related to real-time processing described above have to be solved, since the user's pronunciation will be reproduced in a virtual face and the correct pronunciation will be rendered from plain text.

Besides, the application will be run in a web browser. For that, WebGL has been selected to render the virtual model via web. WebGL [2] is a powerful Application Programming Interface (API) to present 3D graphics in a web page. It is based on OpenGL, which is a very widely used open source 3D graphics standard. Moreover, WebGL is compatible with different and most common browsers such as, Google Chrome, Mozilla Firefox, or Safari. So, WebGL allows the use of web technologies, which are an easy way that non-expert users have to access content.

In the reminder of this chapter, we first make a short review of the state of the art in the Sect. 2; then, in the next two sections, we present the modules for facial animation, together with the mathematical model used, to finally conclude the chapter in the Sect. 5.

2 Related Works

Radovan and Pretorius [3] classify the different methods for facial animation in three groups: based in geometry, based in images and based in real movements.

In the first group, we can find the most primitive works, where transformations of the vertices in the facial mesh are computed one by one. The most relevant method in this group is Parke's [4], since it is considered the first work in facial animation. On the other hand, more recent techniques can also be classified in geometry-based animation. Works that make use of muscles or pseudo-muscles [5] or directly simulate the physical behavior of the muscles and the skin [6] exist nowadays.

Other methods, usually focused on the film industry, are based on images instead of geometry: morphing techniques [7], where different key faces are interpolated to obtain the animation; texture manipulation methods [8], where the changes in the texture of the facial mesh create the animation and works based on blend-shape interpolation [9], where the intermediate frames between two modeled faces are computed.

Nevertheless, recently performance-driven techniques have replaced others because of their realism. Lots of works that capture the movements of an actor and translate them to a virtual face can be found. Beeler et al. [10] analyze the captured data to recognize predefined motions and launch these predefined movements in the virtual face. In the work presented by Arghinenti [11] the captured data is only a layer of the facial animation engine. Other layers cope with expressions, phonemes and muscles. Deng et al. [12] combine a motion capture system with a face scanning system in order to obtain a realistic animation with low-level details, such as wrinkles. Once they obtain the minimum number of key expressions, blend-shape interpolation is used to generate the final animation.

However, as stated before, the motion capture system cannot be used always. And in these cases, another type of facial animation has emerged recently, skeleton-driven facial animation [13–15]. As in corporal animation, the face of the virtual character has a structure of joints and bones attached. The vertices of the facial mesh are transformed in concordance with the skeleton, since they are associated to certain bones. The project SPEEP can be classified in this group.

Unlike in corporal animation, the skeletons differ widely in skeleton-driven facial animation methods. Some have elastic bones, others have rigid bones; some use segments as bones, others use curves as bones; some have tree structure with its node in the neck, others have disjointed groups of bones.

Apart from the facial animation technique, in the project SPEEP we have to decide how the virtual face will pronounce the needed sentences. For example, when pronouncing the phonemes *p* and *a* in different order (*ap* and *pa*) the behavior of the human mouth is not the same, i.e. the "interpolation" between *a* and *p*, and *p* and *a* is not the same. This is due to the so-called co-articulation.

One of the first works in this field and one of the main references is the one presented by Cohen and Massaro [16]. They create exponential functions for facial parameters that rise until the phoneme exact time and fall afterwards. The coarticulation is obtained by the combination of different phonemes and the functions for their associated facial parameters.

Since the first Works in co-articulation were published, several methods have been presented that take speech unities and convert them in a fluent and realistic facial animation: rule-based, Markov models [17], neural networks [18], to name but a few.

To finish this review of the state of the art, it is important to say that not many works in facial animation for WebGL can be found. The work by Benin et al. [19] is an implementation of a WebGL talking head that works with MPEG-4 Facial Animation Parameters (FAPs) standard. And are different lips models, built using 3D kinematics data, which generate the human facial movements. This work is only developed for Apple iOS mobile devices.

There are some other methods for web-based MPEG-4 facial animation that can be more appropriate, including Face Animation Tables (FATs) [20]. FATs allow precise specification of the movements of the vertices assigned to each FAP.

In conclusion, a lot of research has been done to accomplish realistic facial animation, but very few have been developed to be used with the emerging WebGL technologies. Our work goes in this direction.

3 Animation Pipeline

The application developed in the project SPEEP will be used for foreign language pronunciation teaching. On the one hand, the images and the speech of the user will be used for rendering his pronunciation in a virtual character. On the other hand,

Fig. 1 Overview of the SPEEP system

Visual Speech Synthesis, that is a virtual character pronouncing correctly the text, will be used to correct the errors of the user.

The key feature of the project is the utilization of web browser for the rendering of the virtual face. To generate animations in any device, the object has to be drawn many times in a second, drawing it in a different position each time. With this goal, the position of the model should be calculated every time. This means that the animation rate depends on the speed of the rendering cycle. A slower rendering rate would produce choppy animations and a too fast rendering rate would create the illusion of objects jumping from one side to the other. This is the reason why the rendering time is crucial to generate a smooth animation [21].

For this reason, the computational cost when the object is rendering within a browser is an important factor in our project. WebGL technology is slower than native OpenGL because it uses JavaScript for execution. But it is still nearly seven times faster than Flash owing to GPU acceleration. For this reason this project is developed using this technology.

In spite of the good computational cost of WebGL in general, this kind of methods for facial animation can be slower than expected. The speed of the construction of the FATs [20] for example has to be improved to achieve better results.

Thus, our work has been focused on optimizing each step of the facial animation engine, in order to obtain an efficient system. As shown in the Fig. 1, the facial animation system of the project SPEEP is divided in four different parts: the Viseme/Facial Animation Parameter (FAP) Converter, the Co-articulation Module, the FAP/Bone Converter and the Facial Engine. Each part has been or will be studied to reduce the computational cost.

3.1 Viseme/FAP Converter

On the one hand, a viseme is a generic facial image that can be used to describe a particular sound, i.e. it is the equivalent of a phoneme in the field of facial positions. For example, the equivalent of the phoneme *a* is the face that pronounces the phoneme. On the other hand, Facial Animation Parameters (FAPs) were presented in the standard MPEG-4 [22] to define the reproduction of emotions, expressions and speech pronunciation.

The system developed by the project, and presented within this chapter, receives a set of ordered visemes that have to be reproduced and the exact times when they have to be reproduced. So, the first step in the system will be the conversion of these visemes to values of FAPs. The correlation between visemes and FAPs is defined and stored before the real-time performance and when rendering the virtual face, the system only takes the needed values from the stored data and weights them with the input. For example, if the converter receives the viseme *a* with the weight 0.6, the output will be the corresponding FAP's stored values weighted with 0.6.

3.2 Co-articulation Module

The second module of the system receives the values of the FAPs at the moment when the visemes have to be rendered and has to compute the values at any moment. Nowadays, the model of co-articulation is based on Cohen and Massaro's [16] work. Nevertheless, since it is a key aspect for the realism, this is a clear field for our future work.

3.3 FAP/Bone Converter

Once we have the values of FAPs at any time, they must be translated to the facial engine, i.e. they have to be translated to the skeleton that moves the virtual face. A rule-system method is used to convert the FAPs into transformations of the corresponding bones.

For that, a skeleton based on the one defined in the MPEG-4 standard [22] has been defined. In our case, as the system has to work in WebGL, the number of bones has been lowered in order to overcome system's low power. Based on the work by Contreras [23], some sets of bones defined in the standard were replaced by a unique bone, mainly in the lips, ears, nose and jaw. The reduction of the number of bones has been done so that the main areas for the facial animation are not too affected. In the Fig. 2, final bone positions can be seen.

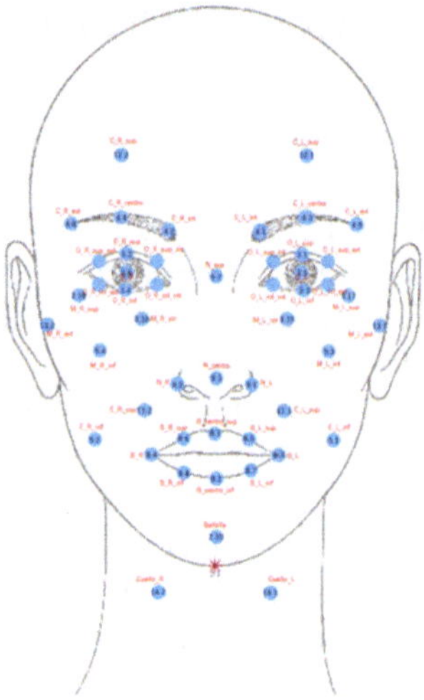

Fig. 2 Positions of the bones defined for project SPEEP

3.4 *Facial Engine*

Finally, the facial engine will receive the transformations that must be applied to the bones of the skeleton and will perform the facial animation. The engine moves the skeleton, and the vertices of the facial mesh are transformed accordingly. In the following section the details of the mathematical model used for the facial animation are shown.

4 Animation Model

During the animation of the virtual face of the project SPEEP, the transformations of the vertices that compound the virtual face are defined by Skeleton Subspace Deformation [24], a widely used and efficient animation technique. The position of the vertex is determined by the transformation of the bones of a skeleton:

$$v = \sum_i w_i T_i \hat{T}_i^{-1} \hat{v} \tag{1}$$

where v is the new position of the vertex, T_i is the transformation of the ith bone in the current pose, $\hat{T}_i^{-1}$ is the inverse of the transformation of the ith bone in the initial pose, $\hat{v}$ is the position of the vertex in the initial pose and w_i is the weight assigned for each bone. The weights w_i must fulfill the following condition:

$$\sum_i w_i = 1 \tag{2}$$

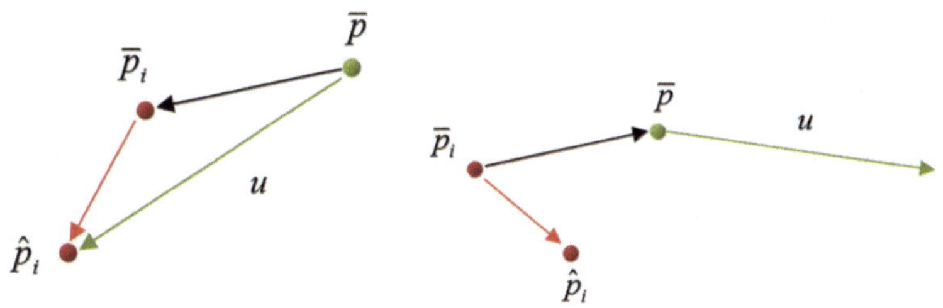

Fig. 3 In the *left*, the movement of the ith bone (*red arrow*) pulls the neighbour bone (*green arrow*). In the *right*, the movement of the ith bone (*red arrow*) pushes the neighbour bone (*green arrow*)

This way, the new position of the vertex is a linear combination of the positions obtained transforming the vertex with the movements of the neighbour bones. For example, the new position of a vertex located in the biceps is obtained by transforming its position with the transformations of the shoulder and the elbow and combining them with their weights.

Nevertheless, the movements obtained by Skeleton Subspace Deformation are very rigid, i.e. they are really appropriate for corporal animations, but too rigid for facial animation. For example, when moving the mouth, only the vertices around it are moved and in order to obtain a high level of realism, it is important that the cheeks and the temples also move.

Thus, in the animation engine, before transforming the vertices, we apply an elastic layer to the bones that allows the system to animate the key part of the animation (the mouth) and automatically animate the other parts (cheeks and temples). A key characteristic of the new system must be its efficiency, since the system works in WebGL.

Therefore, first the system will compute the directions of the translations that will be applied to the bone, i.e. the effects that the bones of the mouth will have on the bones of the cheek:

$$u = \left(\hat{p}_i - \bar{p}_i\right) + \delta_i \left(\bar{p} - \bar{p}_i\right) \tag{3}$$

where u is the direction of the effect, $\hat{p}_i$ is the position after the first transformation, $\bar{p}_i$ is the initial position, i is the index of the neighbour bone and δ_i is the sign of the scalar product of the two vectors in the summation. This way, with the combination of the segment between two bones and the segment between the actual position of the first bone and its initial position, the first bone pulls or pushes the other. The sign of the scalar product is inserted in the summation in so the effect in the bone occurs in the right direction, as shown in the Fig. 3.

Once the directions of the new transformations are obtained, we sum them weighted with the distance between the initial position and the position after the regular transformation of the bones, $\left\| \hat{p}_i - \bar{p}_i \right\|$, and the weight assigned to each bone, w_i. Then, we obtain the last position of the bone, p.

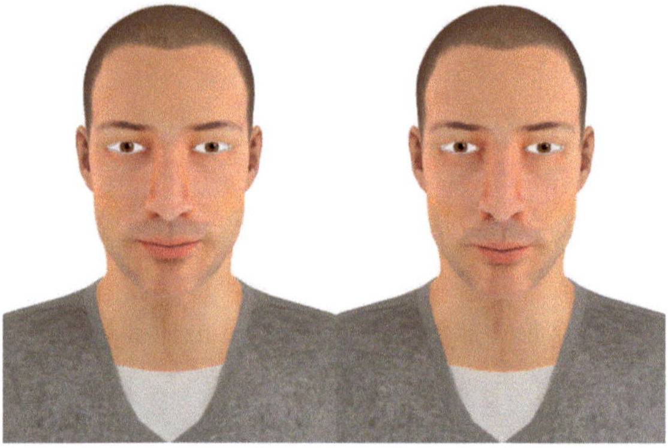

Fig. 4 In the *left*, a frame of an animation without the elastic layer and in the *right*, the same frame with the elastic layer. The cheek of the second figure is translated according to the translation of the mouth

$$p = \hat{p} + \sum_i w_i \frac{u}{\|u\|} \|\hat{p}_i - \overline{p}_i\| \tag{4}$$

This way, the movements of neighbour bones pull a bone, but it is weighted by the assigned weight and the magnitude of the transformation of the neighbour bones. The bigger movement the bone will suffer, the bigger the effect in the neighbor bone will be and, in turn, the further the neighbor will move, the lower will be the effect this will be subject to.

In this moment, the project is focused on the training of this mathematical model of the animation. A facial motion capture system based on the work of Dornaika and Davoine [25] will be used to set the weights of the equation regarding the position of the bones and the predefined rules that relate the values of the FAPs and the rotations of the skeleton.

The first results are very promising, as it can be seen in Fig. 4. As the difference between the animations with and without the elastic layer is not noticeable in one frame, the weights of the elastic layer have been increased to show the difference (note the difference in the cheeks).

5 Conclusion

We have presented the project SPEEP and its facial engine's pipeline. As the project will render a realistic virtual character in real-time by WebGL technologies, it is very important that the animation process is completely optimized.

For that purpose, we have defined a skeleton with a lower number of bones than the MPEG-4 standard without losing the main information of the movements of the face. We have also defined a mathematical model that makes use of the well-known and efficient Skeleton Subspace Deformation and adds a simple elastic layer to obtain a realistic facial animation with a low loss of efficiency in the system.

Although we are in a preliminary phase of the project, the first results have been very promising. Now, we are in the process of training the mathematical model in order to get the most realistic model.

Finally, as stated before, the behaviour of the bones, the simulation of facial muscles, is very important for realism, but the co-articulation, the interpolation between visemes, is even more crucial, since it is useless to have a realistic behaviour in the face if the mouth does not move correctly. So, the main goal in the rest of the project will be the development of an efficient co-articulation algorithm that makes the virtual character realistic enough to read its lips.

References

1. Hodgins, J., Jörg, S., O'Sullivan, C., Park, S.I., Mahler, M.: The saliency of anomalies in animated human characters. ACM Trans. Appl. Percept. **7**(4), 22 (2010)
2. Leung, C., Salga, A.: Enabling WebGL. In: Proceedings of the 19th International Conference on World Wide Web, pp. 1369–1370 (2010)
3. Radovan, M., Pretorius, L.: Facial animation in a nutshell: past, present and future. In: Proceedings of the 2006 Annual Research Conference of the South African institute of Computer Scientists and Information Technologists on IT Research in Developing Countries (2006)
4. Parke, F.I.: Computer generated animation of faces. In: Proceedings of the ACM Annual Conference, vol. 1 (1972)
5. DeRose, T., Kass, M., Truong, T.: Subdivision surfaces in character animation. In: Proceedings of the 25th Annual Conference on Computer Graphics and Interactive Techniques, pp. 85–94, ACM (1998)
6. Yarimizu, H., Ishibashi, Y., Kubo, H., Maejima, A., Morishima, S.: Muscle-based facial animation considering fat layer structure captured by MRI. In: SIGGRAPH'09: Posters, p. 9, ACM (2009)
7. Su, M.-C., Liu, I.-C.: Application of the self-organizing feature map algorithm in facial image morphing. Neural Process. Lett. **14**, 35–47 (2001)
8. Fei, K.: Expressive textures. In: Proceedings of the 1st International Conference on Computer Graphics, Virtual Reality and Visualisation, pp. 137–141, ACM (2001)
9. Huang, H., Chai, J., Tong, X., Wu, H.-T.: Leveraging motion capture and 3D scanning for high-fidelity facial performance acquisition. ACM Trans. Graph. **30**, 74 (2011)
10. Beeler, T., Hahn, F., Bradley, D., Bickel, B., Beardsley, P., Gotsman, C., Sumner, R.W., Gross, M.: High-quality passive facial performance capture using anchor frames. ACM Trans. Graph **30**, 75 (2011)
11. Arghinenti, A.: Animation workflow in KILLZONE3: a fast facial retargeting system for game characters. In: ACM SIGGRAPH 2011 Talks, p. 37, ACM (2011)
12. Deng, Z., Chiang, P.-Y., Fox, P., Neumann, U.: Animating blendshape faces by cross-mapping motion capture data. In: Proceedings of the 2006 Symposium on Interactive 3D Graphics and Games, p. 43–48, ACM (2006)
13. Fathom Studios for Deldo.: http://www.creativecrash.com/maya/tutorials/character/c/facial-animation-rig-for-delgo
14. Talking Heads.: http://www.gamasutra.com/view/feature/3089/talking_heads_facial_animation_in_.php
15. Andy Van Straten.: http://andy-van-straten.com/
16. Cohen, M.M., Massaro, D.W.: Modeling coarticulation in synthetic visual speech. In: Thalmann, N.M. (ed.) Models and Techniques in Computer, Animation, pp. 139–156. Springer, Berlin (1993)

17. Yamamoto, E., Nakamura, S., Shikano, K.: Lip movement synthesis from speech based on hidden Markov models. Speech Commun. **26**, 105–115 (1998)
18. Massaro, D.W., Beskow, J., Cohen, M.M., Fry, C.L., Rodriguez, T. : Picture my voice: audio to visual speech synthesis using artificial neural networks. In: Massaro, D.W. (ed.) Proceedings of AVSP'99: International Conference on Auditory-Visual Speech Processing, pp. 133–138 (1999)
19. Benin, A, Leone, G.R., Cosi P.: Web3D'2012, a 3D talking head for mobile devices based on unofficial iOS WebGl Support. In: Proceedings of the 17th International Conference on 3D Web Technology, pp. 117–120 (2012)
20. Gachery, S., Magnenat-Thalmann, N.: Designing MPEG-4 facial animation tables for web applications. In: Multimedia Modeling Conference, pp. 39–56, Amsterdam, Holanda
21. Cantor D, Jones Br.: WebGL Beginner's Guide. Packt Publishing (2012)
22. Lavagetto, F., Pockaj, R.: The Facial animation engine: towards a high-level interface for the design of MPEG-4 compliant animated faces. IEEE Trans. Circuits Syst. Video Technol. **9**(2), 277–289 (1999)
23. Contreras, V.: Artnatomy. http://www.artnatomia.net/uk/artnatomiaIng.html (2005). Accessed 27 Nov 2012
24. Lewis, J.P., Cordner, M., Fong, N.: Pose space deformation: a unified approach to shape interpolation and skeleton-driven deformation. In: Proceedings of the 27th Annual Conference on Computer Graphics and interactive techniques, pp. 165–172 (2000)
25. Dornaika, F., Davoine, F.: On appearance based face and facial action tracking. IEEE Trans. Circuits Syst Video Technol. **16**(9), 1107–1124 (2006)

Geometric Features for Hand-Written Signatures

Chiara Pellegrini, Anna Rimoldi and Massimiliano Sala

Abstract Hand-written signatures represent nowadays the primary mechanism for authorization in legal transactions. Accordingly, the search for algorithms that allow an effective automatic comparison between images of hand-written signatures turns out to be of the greatest importance. Studies in this area have to face some problems due to intrapersonal variations and interpersonal differences, that make it necessary to analyse signatures as complete images and not as letters and words put together. Therefore, only an extensive analysis of these geometric objects can suggest the most suitable features that have to be evaluated. Such features can be extracted at local or global level of the image and then have to be compared with an appropriate classifier. Moreover, the geometric features often need to be processed on remote clusters, still guaranteeing their privacy. So, we identified some features that could be protected using a suitable form of homomorphic encryption.

Keywords Off-line hand-written signature · Signature verification system · Signature feature extraction · Cryptography · Homomorphic encryption

Mathematics Subject Classification (2010) 68U10 · 11T71

C. Pellegrini · A. Rimoldi · M. Sala (✉)
Department of Mathematics, University of Trento, Via Sommarive 14,
38122 Trento, Italy
e-mail: maxsalacodes@gmail.com

A. Rimoldi
e-mail: anna.rimoldi@gmail.com

C. Pellegrini
e-mail: c.pellegrini88@gmail.com

R. De Amicis and G. Conti (eds.), *Future Vision and Trends on Shapes, Geometry and Algebra*, Springer Proceedings in Mathematics & Statistics 84,
DOI: 10.1007/978-1-4471-6461-6_8,

Fig. 1 An example of interpersonal variations: three different genuine signatures belonging to the same person

1 Introduction

Hand-written signatures represent nowadays the primary mechanism both for authorization and authentication in legal transactions. Accordingly, the search for algorithms that allow an effective automatic comparison between images of hand-written signatures turns out to be of the greatest importance. A signature represents the fruit of a process based on a sequence of actions imprinted in the brain. As a consequence, each signature belonging to the same person can be very variable, depending on the physical and psychological conditions at the moment of acquisition. So intrapersonal variations and interpersonal differences make it necessary to analyze the signatures as complete images and not as letters and words put together. The remainder of this paper is organized as follows:

- A description of the process of signature verification is given in Sect. 2.
- An overview on the features that can be extracted from hand-written signatures is presented in Sect. 3.
- Section 4 provides a summary of the possible classification methods.
- The forgery types and the performance evaluation are described in Sect. 5.
- Section 6 is dedicated to a particular learning approach, known as "writer-independent". In this context, the case of an SVM classifier is analyzed.
- Section 7 looks at a different aspect of the problem, the feature protection. It discusses a possible homomorphic encryption scheme that could be useful for this task.
- Finally, in Sect. 8 our results are presented.

2 The Process of Automatic Signature Verification

The process of automatic signature verification can be subdivided in two steps:

- *learning*;
- *testing*.

The purpose of the learning step is to create a reference file: the input signatures are processed (*pre-processing*) and their personal features are extracted (*feature extraction*) and stored into the knowledge base.

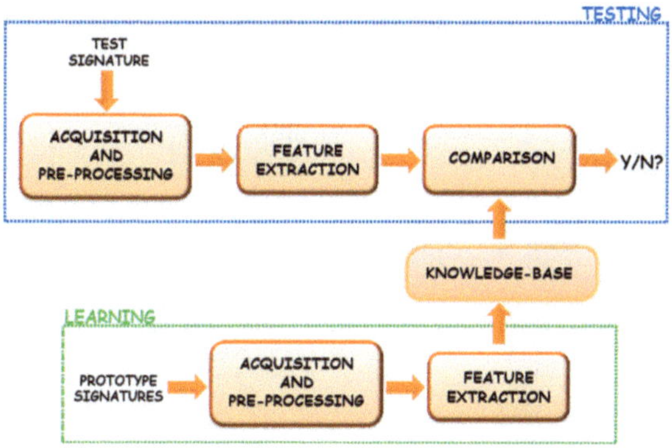

Fig. 2 The process of automatic signature verification

The *pre-processing* step aims to perform an enhancement of the input data, by means of techniques originating from standard signal processing algorithms, such as noise removal, signature size normalization, binarization, rotation and thinning. This is a very sensitive stage, since a wrong choice of the pre-processing operations can involve a great loss of information. In this paper we assume that all the selected features come from binarized and pre-processed images and that they are reduced to a binary (fixed-length) representation.

Note that, in order to extract the aforementioned features, sometimes a segmentation grid could be used. In fact, each feature can be labeled as *global* or *local*, depending on whether it concerns the whole signature or a specific part. A typical global parameter is the global orientation of the signature. In proportion to the level of detail considered, local parameters can be divided into "component-oriented" parameters, which are extracted at the level of each component, and "pixel-oriented" parameters, which are extracted at pixel level. As regards global features, we note two facts. The first is that any distortion of an isolated region will result in small changes on every global feature. The second is that global parameters are less sensitive to signature variation and to the effect of noise. Vice versa, other regions of the signature do not affect local features, and they are sensitive to noise, even in case of small distortions. As a consequence, local features are much more accurate than global features but they are computationally expensive. Features are described in more detail in the next section.

The personal features extracted from an input signature are compared against the information in the knowledge base. This means that, during the testing step, the similarities between the signature under question and its corresponding reference signature are computed, in a way to check whether the tested signature is genuine (*classification*). The purpose is to evaluate the evidence presented in the values of the features obtained from feature extraction and to make a final decision.

Also the testing step requires preprocessing and feature extraction but in addition it needs a classification phase, which will be discussed in Sect. 4.

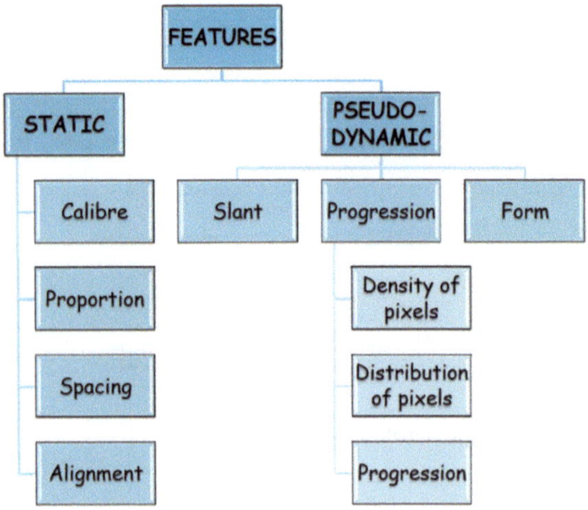

Fig. 3 The two main groups of features: static and pseudo-dynamic

3 Features for Hand-Written Signatures

The crucial step in the signature verification process is the feature extraction, which aims to characterize the signature as a vector of elements, each one representative of the value of a parameter feature. It is preferable to extract those features that will enable the system to correctly discriminate the class of genuine signatures from the classes of forgeries. In particular, there are two main kinds of features that can be extracted: *static* and *pseudo-dynamic* features.

The first group is related to the occupation of the graphical space and includes essentially four types of features [16].

- "*Caliber*" describes the relationship between height and width, as represented in case (a) of Fig. 4. A particular example is the *aspect ratio* (i.e. the ratio of width to height); using such a feature, a possibility consists in classifying signatures into either high-dimension signatures, when the height of the letters is bigger than the width, low-dimension signatures, otherwise.
- "*Proportion*" refers to the regularity and symmetry of the signature. In practice, one checks how the signature is distributed within an imaginary rectangle: if all the letters have more or less the same dimensions and so remain inside such a rectangle (proportionate signature), if some letters protrude from above or below (disproportionate), or if one can distinguish two proportional parts (mixed). See case (b) of Fig. 4.
- "*Spacing*" shows when the writer put pen lifts and breaks between specific letter or stroke combinations. For example, we can distinguish between concentrated signatures (if all the letters are written one attached to the other and the same

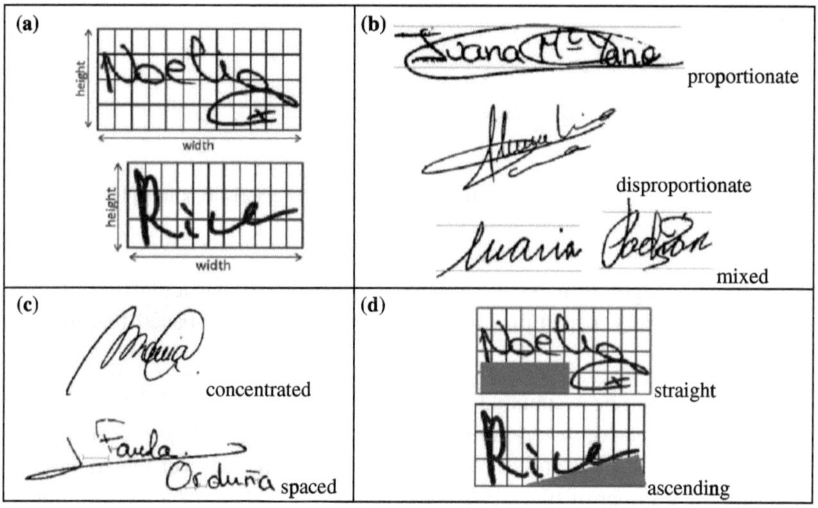

Fig. 4 Some examples of static features: **a** caliber, **b** proportion, **c** spacing, **d** alignment

happens for name and surname) and spaced signatures (if there are spaces between names or between letters into the same word). See case (c) of Fig. 4.

- "*Alignment to baseline*" is simply the relationship of the writing to a (real or imaginary) baseline. In order to identify such a baseline, one can search for an invisible line under the middle zone and at the upper lower zones of the signature. Then, the inclination angle of the baseline is calculated based on a horizontal line that crosses over the baseline. According to this measure, we can finally classify the signature as ascending, straight or descending [27]; an example is given in case (d) of Fig. 4.

Although these features are not so distinctive of the human writing, they usually are tolerant to intrapersonal variation. Moreover, we can observe that features like, for example, alignment work well only if normalization and rotation are correctly performed. This means that there must be also an opportune pre-processing (see Sect. 4).

The second group is much more sensitive, since it is directly related to the strokes of the signature and allows to recover some pressure information from its image. The parameters of this set are mainly geometric-based, i.e. based on the shape and the dimensions of the signature image. In fact, here we can find for example pixel distribution, slant, concavity, number of characteristic points (end points, cross-points, cusps, loops, etc.), contour, direction, curvature and stroke orientation.

It is worth noting that the pseudo-dynamic features are usually of local type, i.e. they are computed in small portions of the signature, defined by a grid. In fact, in this case the signature image is first divided into rectangular regions and then well-defined image characteristics are evaluated in each region.

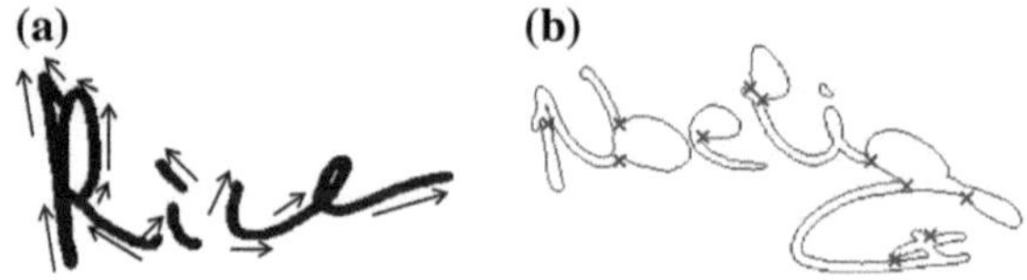

Fig. 5 Two examples of geometric features: **a** slant and **b** characteristic points [1]

Among them, progression, slant and form seem to be particularly suitable for automatic signature verification [16], as well as the contour-based features (more on these later).

- "*Progression*" can be associated to three different values: *pixel density, pixel distribution and effective progression. Pixel density* (or *apparent pressure*) describes the width of the strokes and is computed by placing a grid over the signature and counting the number of black pixels in each cell. *Pixel distribution* computes the width of the stroke in four directions (i.e. in each cell one makes a projection of the strokes according to four zones) [17]. A more complex approach, but based on the same idea, was proposed by Sabourin et al. [24]. *Effective progression* is based on the level of tension in each cell. To be more precise, we select the most significant stroke of each cell, i.e. the longest one, and compute the number of times the stroke changes direction. If few directions are changed, we classify it as a *tense* stroke, otherwise as a *limp* stroke. See case (a) of Fig. 6.
- "*Slant*" can be computed as a combination of both global and local slant measures. It is a characteristic pseudo-dynamic feature, since it is directly related to the signature strokes. It is very robust against forgeries, while on the other hand it would not absorb those signatures with high variability. To compute the slant we have to proceed in two steps: first over the entire image, then for each of the 8×10 cells (vertical $\times$ horizontal) of a segmentation grid (the 8×10 is an example). In this way, one obtains a slant value for each cell, where the final local value is given by the most frequent value in the matrix. This measure of slant angle should reflect both local and global slant characteristics of a signature [21]. Since the predominant axial slant is a value representing the predominant inclination in each cell, for each cell the number of three pixel connections has to be computed against the following templates:
  See case (b) of Fig. 6. (and case a) of Fig. 5.
- "*Form*" stands for the pictorial representation of a letter or writing movement and can be captured by means of concavity (i.e. one extracts concavity measures for each cell of the grid). This is probably the most basic of individual characteristics. See case (c) of Fig. 6.

Other features that are often used in literature are the *contour-based features*. They can be captured at global level, as envelopes extracted at the level of the whole signature, as well as at local level (i.e. at the level of each connected component), according to the classification method that has to be used. For example, working

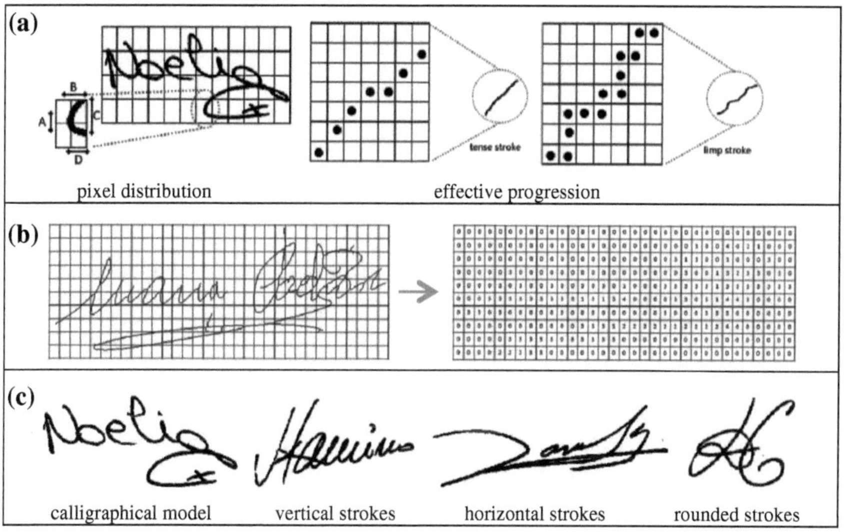

Fig. 6 Some examples of pseudo-dynamic features: **a** progression, **b** slant, **c** form

at local level one can exploit directional properties of signature contours, such as contour-direction, contour-hinge and direction co-occurrence [8]. Otherwise, one can consider only the exterior contour of a signature [4]. In practice, one may extract the exterior contour pixels of all connected components (by means of some image processing algorithms). Then, all the contours in the image are combined into a single contour to obtain a unique closed loop.

4 The Classification Step

In order to decide whether two given signatures match (belong to the same author) or not, we have to perform the classification step. To check whether the tested signature is genuine, the similarities between the signature under question and its corresponding reference signature are computed. Matching the features of the test signature against those stored in the knowledge base, previously developed during the learning step, does this. This process produces a single response (a Boolean value) that states the authenticity of the test signature.
We can identify several approaches to perform the classification, each one heavily dependent on the particular selected features. Such methods can basically be divided in three families:

- *Artificial Intelligence (AI) methods*, such as:
  - Neural Networks (NN) [3], which produce good performance in learning and generalizing;

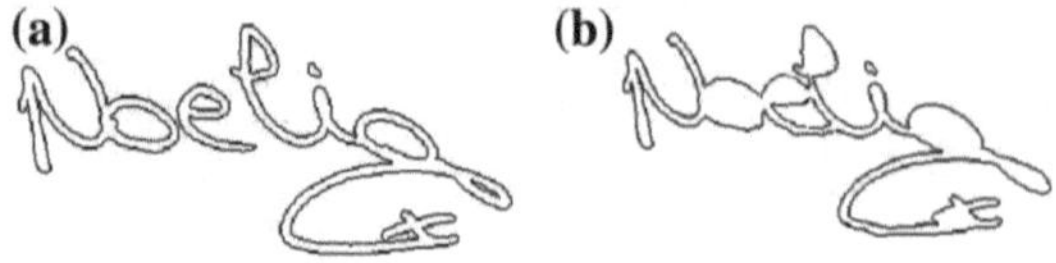

Fig. 7 **a** Counter and **b** exterior conter of a signature

  - Support Vector Machines (SVM) [17, 18], which can map input vectors to a higher dimensional space, where clusters may be determined by a maximal separating hyperplane;

- *Hidden Markov Models (HMM)* [5, 12], i.e. stochastic models which absorb both the variability and the similarity between patterns;
- *Distance-based classifiers*, which usually exploit Euclidean, Hamming or even the Mahalanobis distance. The Mahalanobis distance is based on correlations between variables by which different patterns can be identified and analyzed. It differs from Euclidean distance since it relies on the correlations of the data set. Moreover, it is scale-invariant. See [13].

SVM is a relatively new learning method used for binary classification, developed by Vapnik [6]. The main idea is the following: the input vectors, which could be the feature vectors extracted from the signatures to be compared, are embedded into some high dimensional feature space, through some non-linear mapping ϕ, chosen a priori. In this space, an ad-hoc linear decision surface is constructed. Two problems arise in this process:

- among all hyperplanes separating the learning data, to find the one that generalizes well also for unknown data existing in the testing phase;
- to treat computationally such high-dimensional spaces.

The first problem can be solved if the classes are linearly separable, by means of *optimal hyperplanes*, which are the linear decision functions with maximal margin between the vectors of two classes. However, this can lead to computational problems and overfitting. The second problem can be eliminated by dealing with the high-dimensional space in an indirect way. In fact, knowing the formula (*kernel*) for the dot product in the higher dimensional feature space, it is not necessary to deal with the mapping ϕ directly. In this way, using different dot products, one can construct different learning machines with arbitrary types of decision surfaces. If the classes are not separable, the main idea is to increase the dimension of the feature space, instead of increasing the complexity of the classifier, which remains a hyperplane.

The AI methods and the HMM methods require a lot of computational power and consequently are infeasible if we treat thousands of signatures. On the other hand, the methods of the third class exploit a less sophisticated approach. In fact, it consists in computing the distances among the feature vectors of the genuine signatures of the knowledge base and determining a threshold on the basis of the values obtained. The classification of an input signature successively occurs by checking if its distance

from the prototypes stays below or above the threshold. The threshold is determined from the probability that a given sample belongs to the expected probability distribution. The accuracy of these methods is hampered by the uncertainty in the definition of the conjectured probability distribution. Typically a normal distribution is assumed for simplicity, but sometimes this is not realistic.
We will see an example of such a classifier in Sect. 8.

5 Genuine Signatures Versus Forgeries

The performance of an automatic signature verification system is usually evaluated in terms of two types of error:

- *false rejection rate* (*FRR*), i.e. rejection of genuine signatures;
- *false acceptance rate* (*FAR*), i.e. acceptance of forged signatures.

Naturally, any parameter change in a classification that provides a reduction of the FAR would also be likely to increase the FRR, and vice versa. In general, which type of error is better to minimize depends on the application.

One can also consider measures of the overall error of a system:

- *equal error rate (EER)*, which is obtained when FRR = FAR;
- *total error rate (TER)*, which is defined as $TER = FRR \cdot P(\omega_1) + FAR \cdot P(\omega_2)$, where $P(\omega_1)$ and $P(\omega_2)$ are the a priori probabilities of classes of genuine signatures (ω_1) and forgeries (ω_2).

Note that, since signature forgeries are the results of a behavioral activity, they are strongly related to the type and amount of information provided to the forger, as well as his/her training and effort. Consequently, the FAR evaluation often turns out to be difficult and imprecise. The traditional method of handling this problem consists in assessing three different classes of forgeries:

- *random forgeries*, produced without knowing either the name of the signer or the shape of his/her signature;
- *simple forgeries*, produced knowing the name of the signer but without having an example of his/her signature;
- *skilled forgeries*, produced after training to imitate a genuine instance of the signature as closely as possible.

See Fig. 8 for examples of different forgeries.
It could be unexpected that we need to assess random forgeries. They are possible when the forger hopes that the automatic checks on the signature are not accurate and consequently he/she simply tries with a random signature, for example his/her own name. Naturally, it will be more and more difficult to detect forgeries when passing from random to simple and skilled ones. In particular, each type requires a different recognition approach. In fact, some features or classification techniques, which have proved to be effective in random forgery detection become ineffective in

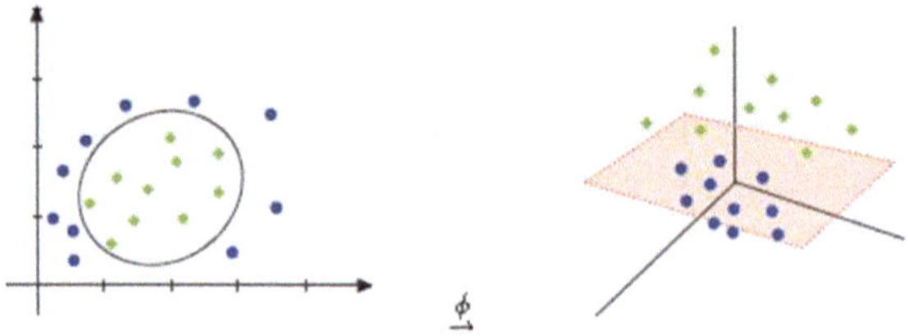

Fig. 8 The two classes, circles and squares, are not linearly separable in the input space, represented on the left. However, by means of the function , the points are mapped in a space, where they become linearly separable

Fig. 9 From *left* to *right*: a genuine signature and three different forgeries of it (in order, random, simple and skilled)

skilled signature verification, because a skilled forgery is usually very similar to the genuine signature. Accordingly, to detect random forgeries one can use global features, because the general aspect of the signature is usually sufficient to discriminate between genuine and forged signatures. On the contrary, in the context of skilled forgeries, it is more suitable to use local features, after a segmentation process of the image [23].

6 The Writer-Independent Approach

Along with the classification technique, special attention is also given to the knowledge-base development in relation to learning strategies.

The most common approach consists in performing a *writer-dependent* learning step, using only genuine specimens. In this case, one learns from samples of a specific individual and uses only these parameters in matching for that individual. Each time a new subject should be included in the system, the main drawback is that an ad-hoc learning step has to be performed. Furthermore, a great number of genuine samples of signatures are necessary to build a reliable model, but in real applications only a limited number of signatures per writer is generally available.

To overcome these problems, a *writer-independent* approach has been proposed. In this case, a set of signature samples is prepared with the help of several individuals and the parameters derived from such a set are exploited in determining whether an arbitrary pair of signatures match or not. Using such an approach, the main challenge of the learning task is to separate the class of genuine signatures and that of forgeries, taking into account that the threshold between them is often not easy to find.

One possibility to implement a writer-independent approach for signature verification consists in exploiting a Support Vector Machine as classifier, as follows.

Given a set of n genuine signature samples as references, $S = \{S_i : i = 1, 2, \ldots, n\}$, we have to compare each S_i with a sample under question Q and to verify the discrepancies among S_i and Q. Therefore, a set of m features is extracted from both the reference image and the one under question. This produces two feature vectors, denoted as $S_i = (v_1(i), \ldots, v_m(i))$ and $V_Q = (v_1(Q), \ldots, v_m(Q))$ respectively. The next step consists in computing the dissimilarity feature vectors $Z_i = |V_i - V_Q|$. After that, the Z_i feed the SVM classifiers C_i, each of which provides a partial decision R_i. The final decision $\mathfrak{D}$ will be obtained through the majority vote rule, based on the fusion of the previously n partial decisions. Note that, in the learning stage, the feature distances Z_i are computed using a pair of signature samples. If the signatures belong to the same writer, the distance feature vector is set to $+1$ (authorship), otherwise it is set to -1 (no authorship). The SVM is then trained to discriminate small feature distances (similar signatures) from large ones (dissimilar signatures). In the classification stage, the image under question is compared with all prototypes through the SVM and this will assign a given distance feature vector to one of the two classes. If the majority of the votes are given to the authorship class, then the system decides that Q is similar to the reference set S, otherwise, it decides that they are dissimilar.

7 An Approach to the Protection of Hand-Written Signatures

The geometric features extracted from the signatures may need to be processed on remote clusters, still guaranteeing their privacy. This is the problem of *template protection*. Since standard encryption techniques are not useful for this task, we exploit *homomorphic encryption*.

A *homomorphic encryption scheme* is a cryptosystem $(\mathcal{P}, \mathcal{C}, \mathcal{K})$ whose encryption function φ_k is a homomorphism and thus preserves group operations performed on the ciphertext set $\mathcal{C}$. This usually allows a third party to take $y_1 = \varphi_k(x_1)$, $y_2 = \varphi_k(x_2) \in \mathcal{C}$ (where $x_1, x_2 \in \mathcal{P}$) and to calculate $\varphi_k(x_1 + x_1)$ or $\varphi_k(x_1 \cdot x_1)$, without knowledge of any secret information. In particular, a *fully homomorphic encryption scheme* (*FHE*) allows the evaluation of arbitrary functions on encrypted data. Such a scheme was shown to exist by Gentry in his breakthrough thesis [7]. Unfortunately, although it would have a myriad of potential applications, it is too expensive for real-life applications, therefore it cannot be used in practice currently. So *somewhat homomorphic schemes* (*SHE*), which support only a limited number of homomorphic operations (below some complexity threshold), have been proposed. These ones seem to be much faster as well more compact.

An interesting SHE scheme is presented [14]. Its main mathematical components, which all depend on a security parameter κ, are: a "dimension" $n = 2^d (d \in \mathbb{N})$, a cyclotomic polynomial $f_{(x)} = x^n + 1$, a prime q such that $q \equiv 1 (mod\ 2n)$, a prime $t < q$, an error parameter $\sigma > 0$ which defines a discrete Gaussian distribution

$\chi = D_{\mathbb{Z}^n,\sigma}$, a bound D on the maximum number of multiplications and a bound A on the maximum number of additions allowed by the scheme. Together, n, q and $f(x)$ define the quotient rings $R = \mathbb{Z}[x]/(f(x))$ and $R_q = R/qR$, while t defines the message space $R_t = \mathbb{Z}_t[x]/(f(x))$.

With all these parameters it is possible to construct a SHE cryptosystem, where the plaintext space is $\mathcal{P} = R_t$, the ciphertext space is $\mathcal{C} = {R_q}^2$ and that works in the following way.

Key generation: Sample a ring element s from χ and define the *private key*

$$s_k = s$$

Sample a uniformly random ring element a_1 from R_q and an error e from χ and compute the *public key*

$$p_k = (a_0, a_1) = (-(a_1 s + te), a_1).$$

Publish p_k, but keep secret s_k.

Encryption: We represent our message $x \in \mathcal{P}$ as a polynomial of degree at most $n-1$ with coefficients in $\mathbb{Z}_t$. Then, we sample u, f, g from χ and we compute the ciphertext $y \in {R_q}^2$ as

$$y = \varphi_{Pk}(x) = (y_0, y_1) = (a_0 u + tg + x, a_1 u + tf).$$

Homomorphic addition: Let $y = (y_0, y_1, \ldots, y_\delta) \in {R_q}^\delta$ and $y' = (y'_0, y'_1, \ldots, y'_\gamma) \in {R_q}^\gamma$ be two ciphertexts. Assume $\delta = \gamma$, otherwise pad the shorter ciphertext with zeroes. Compute and output

$$y_{aad} = (y_0 + y'_0, y_1 + y'_1, \ldots, y_{\max}(\delta, \gamma) + y_{\max}(\delta, \gamma)') \in {R_q}^{\max(\delta,\gamma)}.$$

This means that homomorphic addition is done by a simple component-wise addition of the ciphertexts.

Homomorphic multiplication: Let $y = (y_0, y_1, \ldots, y_\epsilon) \in {R_q}^\delta$ and $y_{'} = (y_0', y_1', \ldots, y_\gamma') \in {R_q}^\gamma$ be two ciphertexts. Here, we do not pad either of the ciphertexts with zeroes. Let v be a symbolic variable and consider the expression (over R_q)

$$\left(\sum_{i=0}^{\delta} y_i v^i\right) \cdot \left(\sum_{i=0}^{\gamma} y'_i v^i\right).$$

Treating v as an unknown variable, we can compute $y_0, \ldots, y_{\delta+\gamma} \in R_q$ such that for all $v \in R_q$

$$\left(\sum_{i=0}^{\delta} y_i v^i\right) \cdot \left(\sum_{i=0}^{\gamma} y'_i v'\right) = \sum_{i=0}^{\delta+\gamma} y'_i v^i.$$

The output ciphertext is $y_{mlt} = (\bar{y}_0, \dots, \bar{y}_{\delta+\gamma}) \in {R_q}^{\delta+\gamma}$.

Decryption: Given the ciphertext $y = (y_0, y_1, \dots, y_\delta) \in {R_q}^{\delta}$, we compute

$$\hat{x} = \sum_{i=0}^{\delta} y_i s^i \quad \in R_q$$

and output the message $\psi_{sk}(y) = \hat{x}(mod t) \in \mathcal{P}$.
Note that the homomorphic operations (in particular, multiplication) increase the number of ring elements in the ciphertext. Moreover, Naehrig's main theorem gives a bound on D and A, depending on the other parameters:

Theorem *Assuming that the following relation holds*

$$q \leq 4 \cdot (2t\sigma\, 2\sqrt{n})^{D+1} \cdot (2n)^{D/2} \cdot \sqrt{A},$$

the previous encryption scheme is correct and can compute D multiplications followed by A additions.

8 Our Contribution

In Sect. 3, we describe some commonly used geometric features that are extracted from hand-written signatures. Now we delve into those that should be most suitable for use in a verification algorithm that can be efficiently implemented using Naehrig's SHE scheme. Naturally, in our selection we checked that the resulting error rates were comparable with those obtained by other methods present in literature. In particular, we consider two types of features: a *slant* distribution and a set of eleven *global features*. We now present the distance-based classifier that we used.

8.1 Our Implementation of a Distance-Based Classifier

The basic idea of this writer-dependent approach consists in analyzing the feature vector components of the prototypes, in order to reduce the variability of the genuine signatures.

For each user, let us consider 10 genuine signatures as prototypes and let us denote by $\mathcal{P} = (f^{(1)}, \dots, f^{(10)})$ the set containing their feature vectors $f^{(i)} = (f_1^{(i)}, \dots, f_m^{(i)})$. We dispose them in a $m \times 10$-matrix as follows:

$$\begin{pmatrix} f_1{}^{(1)} \dots f_1{}^{(i)} \dots f_1{}^{(10)} \\ \vdots \quad\quad \vdots \quad\quad \vdots \\ f_j{}^{(1)} \dots f_j{}^{(i)} \dots f_j{}^{(10)} \\ \vdots \quad\quad \vdots \quad\quad \vdots \\ f_m{}^{(1)} \dots f_m{}^{(i)} \dots f_m{}^{(10)} \end{pmatrix}.$$

Then, we choose the "right" rows, depending on the standard deviation values of each row, in order to obtain an $m^{'} \times 10$-matrix, where $m^{'} \leq m$:

$$\begin{pmatrix} f_1{}^{(1)} \dots f_1{}^{(i)} \dots f_1{}^{(10)} \\ \vdots \quad\quad \vdots \quad\quad \vdots \\ f_{m'}{}^{(1)} \dots f_{m'}{}^{(i)} \dots f_{m'}{}^{(10)} \end{pmatrix}.$$

In practice, the components having a standard deviation σ greater than a pre-fixed threshold σ_{th} are removed from the vector. Notice that the threshold σ_{th} is strongly dependent on the database. In this way, we obtain vectors with a writer-dependent number of components.

Let us denote by $\mathcal{P}' = (f'^{(1)}, \dots, f'^{(10)})$, with $f'^{(i)} = (f_1{}^{(i)}, \dots, f_{m'}{}^{(i)})$, the set of the prototype feature vectors after the component selection. The successive step consists of eliminating (or at least reducing) the influence of the prototypes (or some of their parts) with high difference from the others. To obtain this, we make a selection on the columns, instead of on the rows. That is, we mark the "bad values" of each row of the $m^{'} \times 10$-matrix, obtained with the previous step, i.e. the values outside the range $[\bar{x} - \sigma, \bar{x} + \sigma]$ (where $\bar{x}$ is the mean of the ten values of each component). Then, we compute the mean $\bar{x}_{marked}$ and the standard deviation σ_{marked} of the number of marked values for each prototype and we remove the columns where the number of marked values is greater than $\bar{x}_{marked} + \sigma_{marked}$. In this way, we now consider an $m^{'} \times p$-matrix, where $p \leq 10$ is the number of "survived" prototypes $\mathcal{P}^{''} = (f'^{(1)}, \dots, f'^{(p)})$:

$$\begin{pmatrix} f_1{}^{(1)} \dots f_1{}^{(p)} \\ \vdots \quad\quad \vdots \\ f_{m'}{}^{(1)} \dots f_{m'}{}^{(p)} \end{pmatrix}.$$

Finally, from the resulting matrix, we determine an "average feature vector" F corresponding to the prototypes. Then, we compute the distances of the $\mathcal{P}^{''}$ elements from F and the maximum of these distances is imposed as threshold t. To verify a signature under question, we compute its distance d_Q from the average feature vector and check if it is less or greater than t:

$$\begin{array}{l} if \;\; d_Q \leq t \Rightarrow genuine \\ if \;\; d_Q > t \Rightarrow forgery. \end{array}$$

8.2 Some Results

The results reported below were obtained using two different databases.

- *SVC20* [29] is a subset of the *SVC 2004* database, containing 20 genuine signatures and 20 forged signatures for 20 signers. The images of the signatures were reconstructed based on the data of the on-line corpus.
- *GPDS39* is a subset of the larger database *GPDS960signature* [2], which has often been used for experiments of off-line signature verification. It contains data from 960 individuals: 24 genuine signatures for each individual, plus 30 forgeries of his/her signature. In particular, *GPDS39* contains only the signatures of the first 39 writers of the *GPDS960*signature.

The first feature that we used with the classifier presented above is the slant. For it we use the *Hamming distance* as classifier, with the procedure described above. We exploit this distance measure (and not for example the Euclidean distance) because the values of the slant are simply "labels", they do not have a quantitative meaning. Hence, it is sufficient to check how many components are equal and how many differ. To perform the classification, we made the threshold σ_{th} vary and we obtained the following results:

Database	σ_{th}	FAR_random	FAR_skilled	FRR
SVC20	17	0.007576	0.125000	0.300000
SVC20	18	0.015152	0.141667	0.258333
SVC20	19	0.015152	0.158333	0.258333
GPDS39	24	0.001976	0.089855	0.447205
GPDS39	26	0.001976	0.076812	0.459627
GPDS39	27	0.001976	0.076812	0.459627

As the table shows, the FRR is quite high (almost always above 40 %), while the *FAR* values are maintained at an acceptable level. In most cases, the *FAR_random* is even below 1 %, whatever σ_{th} we choose.

If we also want to protect these features, we need $A = 500$ as bound on the number of allowed additions and $D = 1$ since we do not compute any multiplication. In fact, we work with feature vectors of 240 bits and we have to compute the addition of two of such vectors (component by component) and then sum all bits of the resulting vector. According to Naehrig's theorem, we can perform this matching with their SHE scheme by using keys with lengths comparable to those of elliptic curve encryptions. The operations necessary are many more, but within the capabilities of today's servers.

In the second case, we use a set of eleven global geometric features $f = (NA, AR, \#p_c, C, \#p_R, R, \hat{x}, \hat{y}, NA_l, NA_c, NA_r)$, which can easily be derived from the signature itself [26]. These are listed below:

- *Normalized area of the signature* (*NA*): the ratio of the area occupied by the signature to the area occupied by the bounding box (number of pixels of the box). This means that

$$NA = \frac{A_{signature}}{A_{box}} = \frac{\Sigma_x \Sigma_y b(x, y)}{H_{box} \cdot W_{box}},$$

where $b(x, y)$ is the characteristic function of the signature in the image, H_{box} and W_{box} are respectively height and width of the box. In particular, *NA* provides information about the pixel density.

- *Aspect ratio* (*AR*): the ratio of the width to the height of the signature. This is calculated because width and height of a person's signature may vary, while their ratios will be approximately equal.
- *Centre of area* $(\hat{x}, \hat{y})$: centre of mass of a figure of the same shape, with constant mass per unit area. Its coordinates can be computed as

$$\hat{x} = \frac{1}{A_{signature}} \sum_x \sum_y x b(x, y),$$
$$\hat{y} = \frac{1}{A_{signature}} \sum_x \sum_y y b(x, y).$$

- *Maximum horizontal and vertical histograms* $(\#p_C, C, \#p_R, R)$: they are calculated by going through each row (resp. column) in the signature and counting the pixels in the row (resp. column). The row R (resp. column C) with the highest number of pixels $\#p$ is the maximum horizontal (resp. vertical) histogram.
- *Trisurface feature* (NA_l, NA_c, NA_r): we split the image into three equal parts (left, centre, right) and compute the normalized signature area for each of the three parts.

For the global features we choose the *Euclidean distance* as classifier. Note that the scale of individual features, extracted by the different feature extraction techniques implemented, is completely different. Hence, to perform the classification, a statistical method should be first used to normalize f. So, each component f_i is independently transformed into its normalized value as follows:

$$\bar{f_i} = \frac{f_i - \mu}{\sigma},$$

where μ and σ are mean and standard deviation of f coordinates respectively. Then, the same procedure previously explained yields the following results:

Database	σ_{th}	FAR_random	FAR_skilled	FRR
GPDS39	0.4	0.144269	0.360870	0.263975
GPDS39	0.5	0.144269	0.376812	0.251553
GPDS39	0.6	0.175889	0.392754	0.254658

Instead, without performing any normalization and skipping the first component selection (i.e. we choose $\sigma_{th} = 1{,}000$ in such a way that after the first reduction, the resulting matrix still has dimension $m \times 10$), the errors obtained are:

Database	FAR_random	FAR_skilled	FRR
GPDS39	0.268775	0.550725	0.220497

In this way, the *FRR* is lower, but the *FAR* has still increased.

Here, to use Naehrig's SHE scheme, we need $A = 21$ and $D = 12$. In fact, we use vectors of $m = 11$ components and we have to compute overall $2m - 1$ additions and m multiplications (since it is not possible to compute divisions and square roots homomorphically, these operations have to be performed after decryption). This translates in considering keys with 600 bits. Since these keys guarantee the security of a 1024-bit RSA, the public key length is appropriate, but the operations involved are beyond the capabilities of today's servers. However, there seems to be space for computational refinement in the homomorphic operations, which should lead to practical computations.

Acknowledgments Part of this work was carried out during a project in collaboration with Corvallis SPA and the Free University of Bozen-Bolzano. Other people in the University of Trento took also part in the projects. Therefore, we would like to thank: Lorenzo Nicolodi, Stefan Unterweger and Alex Tomasi.

References

1. Bansal, A., Nemmikanti, P., Kumar, P.: Offline signature verification using critical region matching. In: IEEE Second International Conference on Future Generation Communication and Networking Symposia, 2008 (FGCNS'08), vol. 3, pp. 115–120 (2008)
2. Blumenstein, M., Ferrer, M.A., Vargas, J.F.: The 4nsigcomp2010 offline signature verification competition: scenario 2. In: IEEE 2010 International Conference on Frontiers in Handwriting Recognition (ICFHR), pp. 721–726 (2010)
3. Burr, D.J.: Experiments of neural net recognition of spoken and written text. IEEE Trans. Acoust. Speech Sig. Process. **36**(7), 1162–1168 (1988)
4. Chen, S., Srihari, S.: Use of exterior contours and shape features in off-line signature verification. In: IEEE Proceedings of Eighth International Conference on Document Analysis and Recognition, 2005, pp. 1280–1284 (2005)
5. Coetzer, J., Herbst, B.M., Du Preez, J.A.: Offline signature verification using the discrete radon transform and a hidden markov model. EURASIP J. Appl. Signal Process. **2004**, 559–571 (2004)
6. Cortes C, Vapnik, V.: Support-vector networks. Mach. Learn. **20**(3), 273–297 (1995)
7. Gentry, C.: A fully homomorphic encryption scheme. Ph.D. thesis, Stanford University (2009)
8. Gilperez, A., Alonso-Fernandez, F., Pecharroman, S., Fierrez, J., Ortega-Garcia, J.: Off-line signature verification using contour features. In: Proceedings of ICFHR (2008)
9. Guo, J.K., Doermann, D., Rosenfield, A.: Off-line skilled forgery detection using stroke and sub-stroke properties. In: IEEE Proceedings of 15th International Conference on Pattern Recognition, 2000, vol. 2, pp. 355–358 (2000).

10. Hou, W., Ye, X., Wang, K.: A survey of off-line signature verification. In: IEEE Proceedings of 2004 International Conference on Intelligent Mechatronics and Automation, 2004, pp. 536–541 (2004)
11. Impedovo, D., Pirlo, G.: Automatic signature verification: the state of the art. Syst. Man Cybern. Part C Appl.Rev. IEEE Trans. **38**(5), 609–635 (2008)
12. Justino, E.J.R., El Yacoubi, A., Bortolozzi, F., Sabourin, R.: An offline signature verification system using hmm and graphometric features. In: Fourth IAPR International Workshop on Document Analysis Systems (DAS), Rio de, Citeseer, pp. 211–222 (2000)
13. Mahalanobis, P.C.: On the generalized distance in statistics. Proceeding of the National Institute of Sciences of India **2**(1), 49–55 (1936)
14. Naehrig, M., Lauter, K., Vaikuntanathan, V.: Can homomorphic encryption be practical? In: Proceedings of the 3rd ACM Workshop on Cloud Computing Security Workshop, ACM, pp. 113–124 (2011)
15. Nguyen, V., Blumenstein, M., Leedham, G.: Global features for the off-line signature verification problem. In: IEEE 10th International Conference on Document Analysis and Recognition, 2009 (ICDAR'09), pp. 1300–1304 (2009)
16. Oliveira, L.S., Justino, E., Freitas, C., Sabourin, R.: The graphology applied to signature verification. In: 12th Conference of the International Graphonomics Society, pp. 286–290 (2005)
17. Oliveira, L.S., Justino, E., Sabourin, R.: Off-line signature verification using writer-independent approach. In: IEEE International Joint Conference on Neural networks, 2007 (IJCNN 2007), pp. 2539–2544 (2007)
18. Özgündüz, E., Sentürk, T., Karslıgil, M.E.: Off-line signature verification and recognition by support vector machine. In: European Signal Processing Conference, Turkey, (2005)
19. Pal, S., Blumenstein, M., Pal, U.: Off-line signature verification systems: a survey. In: Proceedings of the International Conference and Workshop on Emerging Trends in Technology, ACM, pp. 652–657 (2011)
20. Pal, S., Blumenstein, M., Pal, U.: Off-line signature verification systems: a survey. In: Proceedings of the International Conference and Workshop on Emerging Trends in Technology, ACM, pp. 652–657 (2011)
21. Qi, Y., Hunt, B.R.: A multiresolution approach to computer verification of handwritten signatures. IEEE Trans. Image Process. **4**(6), 870–874 (1995)
22. Sabourin, R., Cheriet, M., Genest, G.: An extended-shadow-code based approach for off-line signature verification. In: IEEE Proceedings of the Second International Conference on Document Analysis and Recognition, 1993, pp. 1–5 (1993)
23. Sabourin, R., Drouhard, J.P., Wah, E.S.: Shape matrices as a mixed shape factor for off-line signature verification. In: IEEE Proceedings of the Fourth International Conference on Document Analysis and Recognition, 1997, vol. 2, pp. 661-665 (1997)
24. Sabourin, R., Genest G., Preteux, F.: Off-line signature verification by local granulometric size distributions. In: IEEE Transaction on PAMI, 1997, pp. 976–988
25. Santos, C.R., Bortolozzi, F., Oliveira, L.S., Justino, E.: Off-line signature verification based on forensic questioned document examination approach. In: Proceedings of the 2007 ACM symposium on Applied computing, ACM, pp. 637–638 (2007)
26. Schafer, B., Viriri, S.: An off-line signature verification system. In: 2009 IEEE International Conference on Signal and Image Processing Applications (ICSIPA), IEEE, pp. 95–100 (2009)
27. Shamsuddin, M.R., Mohamed, A.: Baseline image classification approach using local minima selection. In: Visual Informatics: Bridging Research and Practice, Lecture Notes in Computer Science, vol. 5857, pp. 727–737 (2009)
28. Srihari, S.N., Xu, A., Kalera, M.K.: Learning strategies and classification methods for off-line signature verification. In: IEEE Ninth International Workshop on Frontiers in Handwriting Recognition, 2004 (IWFHR-9 2004), pp. 161–166 (2004)
29. Xiong, Y., George, S., Kashi, R., Matsumoto, T., Rigol, G., Yeung, D., Chang, H.: SVC2004: first international signature verification competition. In: Proceedings of ICBA 2004, Hong Kong (2004)

Using 2D Contours to Model Metal Sheets in Industrial Machining Processes

Aitor Moreno, Álvaro Segura, Harbil Arregui, Jorge Posada, Álvaro Ruíz de Infante and Natxo Canto

Abstract The cutting and punching industrial processes can be conceptualized as material removal methods, where the sheet is being transformed as the running CNC program is being executed by the NC machine. The simulation of such processes has to take into account this increasing complexity of the metal sheet. In this work we introduce a representation based on 2D contours to model the metal sheet in such industrial processes. All the operations of the programs are transformed to Boolean operations between the sheet and the sweep of moving cylinder (cutting machines) or a complex polygon (for punching machines). As a direct application of the technique results in poor performance, a spatial subdivision has been used to increase the performance.

1 Introduction

In the machining industry, technological advances have led to obtain more efficient machinery, faster and more versatile, leading to better economical results. In the field of metal sheet cutting, the nature of the involved processes generates lot of wasted material, having to be returned to the melting industry for recycling.

Therefore, simulation tools can help to test and check the CNC programs in the design phase, being verified all the necessary times before they are actually run in the machines. After a series of tests, including the optimization modifications, the

A. Moreno (✉) · Á. Segura, H. Arregui · J. Posada
Vicomtech-IK4, Paseo Mikeletegui 57, 20009 San Sebastián, Spain
e-mail: amoreno@vicomtech.org

J. Posada
e-mail: jposada@vicomtech.org

Á. Ruíz de Infante · N. Canto
Lantek Sheet Metal Solutions, Parque Tecnológico de Alava, Albert Einstein 36, Miñano, 01510 Vitoria, Spain

R. De Amicis and G. Conti (eds.), *Future Vision and Trends on Shapes, Geometry and Algebra*, Springer Proceedings in Mathematics & Statistics 84,
DOI: 10.1007/978-1-4471-6461-6_9, © Springer-Verlag London 2014

program can reach an optimized state, good enough to be transferred to production. Any not fully tested CNC program can cause or increase the risks and compromise the machinery, provoking partial breakages of parts, collisions between different machine parts, the metal sheet and the tool.

This work present some of challenges related to the representation of the metal sheet, found during the development of an interactive NC simulator for cutting and punching processes for metal sheets.

The cutting processes operate on the sheet, following the instructions provided by the CNC programs. They are composed of a large number of linear and arc movements, and other auxiliary instructions. All the machining operations can be conceptualized as Boolean operations over the sheet. A linear or arc movement removes the material from the sheet following the path of the torch. Hence, a movement can be conceptualized as a sweep of a moving cylindrical tool.

In the punching processes, the material removed from the sheet is not a sweep of the tool, as a punch instruction only uses its profile to remove material only once. The challenges of the punching processes are related to the complexity of those patterns and the high frequency of such operations (up to 1,000 punches per minute).

2 Related Works

The NC machining simulation using Computer Graphics techniques is a widely extended research topic, where the main issue is related to the representation of the dynamic parts of the simulation (in this case, the metal sheet is considered dynamic, as its geometry changes over time). Some traditional approaches do not store the geometrical information during the simulation, but they simply modify the drawing screen using an image-based approach.

Some techniques store the intermediate results in the computer's memory, having an internal 3D geometric representation of the object that is changed continuously during the simulation process. With these methods, a permanent representation is always available and provides free camera movement around the object, better geometric accuracy control, geometric based collision detection, etc.

Van Hook [3] used an extended Z-buffer data structure (called a Dexel structure) for the graphical verification. In his work, a scan method to convert surface data into his Dexel (depth element) structure was presented. The Z values for the nearest and the farthest surface at each Dexel are stored in such depth elements. This technique has been extended by several authors [12].

Other representation methods in the Computer Graphics field are fundamentally geometric like (1) Boundary Representation (B-Rep), (2) Constructive Solid Geometry (CSG), and (3) Hierarchical Space Decomposition (HSP).

Although B-Rep is the most used method for solid modelling in modern CAD systems, its straightforward use for machining simulation is not convenient due to the long time required for the dynamic simulations [8]. A similar problem occurs with

CSG representation, with computational costs of order $O(n^2)$, where n is the number of primitives [9] being computationally expensive. A modern implementation of CSG representation through BSP (Binary Space Partitioning) trees has been ported to Javascript and to the Web with a great but non real time performance [11].

To cope with the complexity of the problem and the long time required in these approaches, the approximation of the exact geometry, and especially the partitioning of the object in suitable regions has been proposed by several authors [9].

The most classic technique for volume partitioning is the voxel representation (classical octree, extended octree [1], SP-Octree [2]) that combines the space partitioning, solid representation and Boolean operation support in a single definition. The sheet to be manufactured can be approximated by a very thin extruded plane, given that the machining program is limited to 2D movement over the sheet. This sheet representation provides a direct way to perform Boolean operations between the moving tools and the planar sheet. The 3D Boolean operation is simplified in a single 2D Boolean operation between two 2D complex polygons, that is a well-reviewed research topics [7, 10].

In this work, we present how an efficient and optimized metal sheet-machining simulator has been designed and developed, with an internal geometrical core based on complex polygon Boolean operations to support the material removal processes.

3 Punching and Cutting Simulator

A simulation system for cutting and punching of metal sheet through a NC controller was developed. The simulator is as a prototype simulation software module and it is focused to obtain consistent geometric results and high graphics quality for the following sheet machining processes (see Fig. 1):

- Metal sheet cutting processes using laser, plasma, oxy-fuel or water jet.
- Punching processes with tools defined with basic or complex shapes.

By means of simulation techniques, the application emulates the behaviour of the machine tool to the computer. The simulation system takes as input a starting NC program (normally, G-code dialect) translated to a common and simplified XML format, listing all the movements that the machine will perform during the cutting or punching operation. The operation mechanisms differ between the cutting and punching process. The main significant difference is that in the cutting process the removal process is continuous while the torch is powered on, whereas in the punching process, the removal process is instantaneous when the punch is triggered.

In the following subsections, the system architecture and the main functionality will be described.

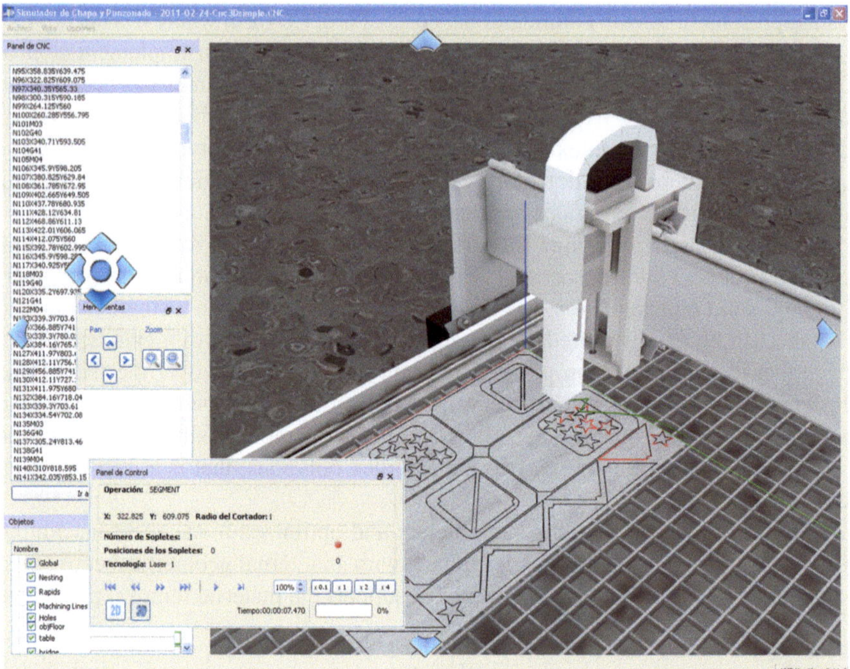

Fig. 1 User interface of the CNC simulator. The UI shows all the necessary information to the user (CNC instructions, simulation controls, visibility options and the 3D interactive output)

3.1 System Architecture

The simulator is structured as a multi-layered architecture, each one encapsulating different methods and techniques. The low level layer involves the geometric calculations (Geometric Kernel) with the management of the Boolean operations between 2D polygons as its main responsibility. The Clipper library [4] provides the functionality to calculate Boolean operations between 2D polygons, as the basis for subtraction between a complex polygon representing the sheet and another polygon representing the sweep of a moving tool in a given period of time (either cutting or punching).

The low level layer provides access to the graphic system, responsible for rendering the simulation results in the screen. OpenSceneGraph [6] was chosen to be our graphic engine, which is used to draw on screen the result of the Boolean operations, the 3D machine models and all the virtual elements in the scene.

The middle layer of the architecture provides the logic behind the application domain. The concepts related to the NC machine domain, like *Part*, *Tool*, *Machine* and *Axis* are represented as classes. The animated behaviour of the moving elements is also represented. In this layer the geometric sweep volume of the moving tools are calculated and passed to the lower level for the actual Boolean operation with the metal sheet representation.

Over the cutting and animation layer, we have added the user oriented layer, which includes the graphical interface of the prototype application and management of the XML files. This has been necessary to:

- Provide the user interface, including multi-language interface options.
- Implement the different navigation methods in the 3D virtual world, using the mouse as input device.
- Manage and display the real NC instructions (highlighting the currently executing instruction) that will be loaded into the simulator through the intermediate and abstract XML representation.
- Manage and display the visualization options of the 3D objects corresponding to the individual parts of the cutting or punching machine.

Both the user interface and the XML parsing functionality have been implemented through the Qt library.

3.2 User Interface

The user interface (see Fig. 1) displays a simple and clean interface between the user and the simulation functionality. It has been developed using the Qt library, which provides the technology needed to create professional and high-quality graphical interface controls. The application GUI is structured as a set of docking and floating panels, which can be moved freely or docked to any side of the screen with the 3D virtual world always in the central widget.

There are two ways of controlling the virtual simulation. One is by running a continuous simulation, where an animation factor is applied to control the simulation speed. The other is a fast mode, which will run the simulation in background as fast as possible, until the target instruction is reached. Combining both modes, users can go to a specific instruction of the NC code (shown in an independent panel) and from there, begin an animated simulation with the desired speed factor. The rest of the buttons in the Control Panel offer simple VCR functionality: play and stop, play only one instruction, and go directly to the previous or next instruction, the first or the last instruction of the loaded program.

The GUI provides the visibility options for all the objects in the virtual scene. Any object can toggle its visibility, but not all the elements can modify the transparency, as the lines (machining toolpaths and other helping elements) and the ground model. The camera properties and movements can be set up in another panel, including the toggle button to go to 3D or 2D mode and the zoom and pan functionality.

Finally, thanks to the multilingual support from Qt, the interface has been easily translated to several languages, with an easy and portable mechanism to add new languages.

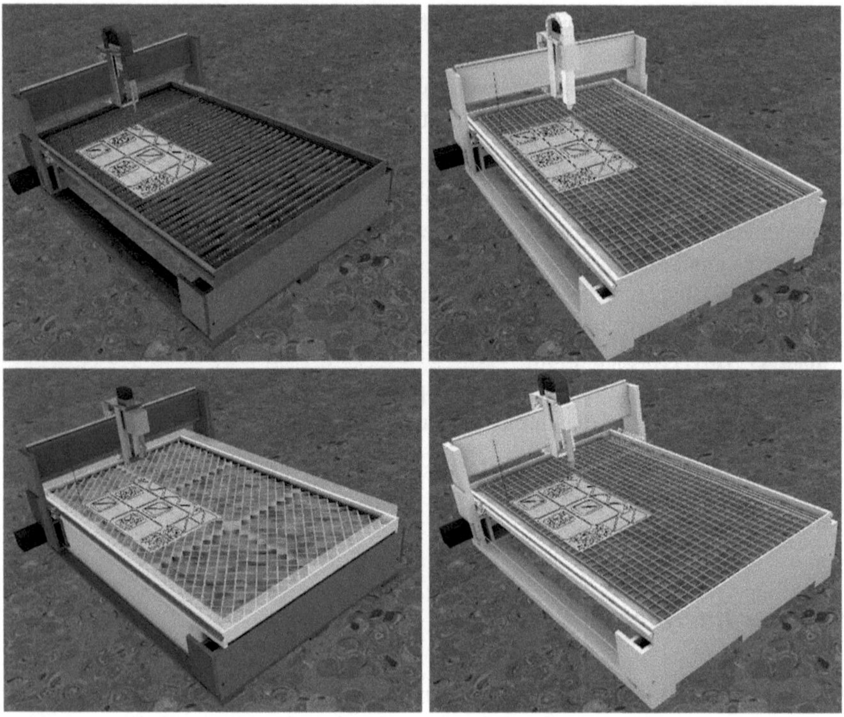

Fig. 2 A specialized cutting machine model has been provided for each technology: oxy-fuel, laser, plasma and water-jet

3.3 Machine Representation

Each cutting process has its own 3D model of the cutting machine. Although all of them behave essentially in the same way, the visual differences are important for the users. In the Fig. 2, the different machines are shown from a similar point of view.

They are composed of a stationary table, a moving bridge over the table (*X axis*) and a moving tool-support over the bridge (*Y axis*). The tool-support contains the tool, which can be replaced with different torches depending on the ongoing work.

The bed is the most noticeable difference in the table. For example, the water jet machine provides a glass frame to avoid water splattering, and the bed's floor is solid. In the other machine, the bed's floor differs, as they are oriented to different materials and thickness of metal sheets.

Each machine has also a different torch, which provides the cutting technology. In the Fig. 3, the torches for oxy-fuel, plasma, water-jet and laser technologies are shown.

The hierarchical structure of the punching machine is significantly different from the cutting machine. In the punching machine, the sheet is the mobile element, while in the cutting machine the sheet is static. Therefore, the sheet is moved in the X and

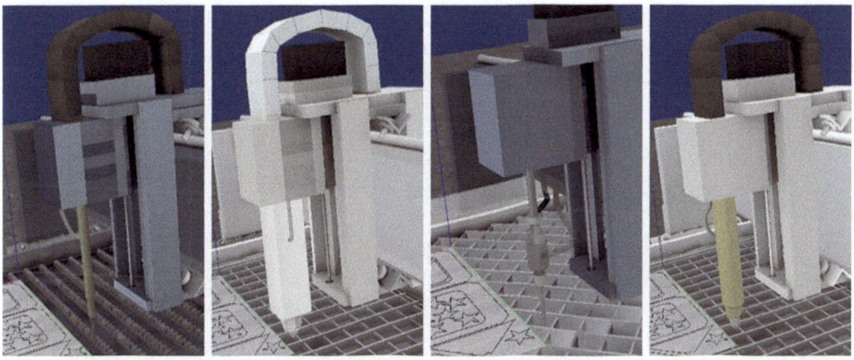

Fig. 3 Each machine has its own model to represent the torches

Fig. 4 Punching machine 3D model. The bridge is shown with transparency to allow users to view the punching processes

Y axis, while the punching stations give the punch movements in the Z axis. The implemented prototype of the punching machine has room for up to 20 individual punching tools. The individual punch geometries are loaded from an XML file.

The visualization of the punching machines requires playing with transparencies in the punching station and the bridge, since they hide completely the punching operation (see Fig. 4).

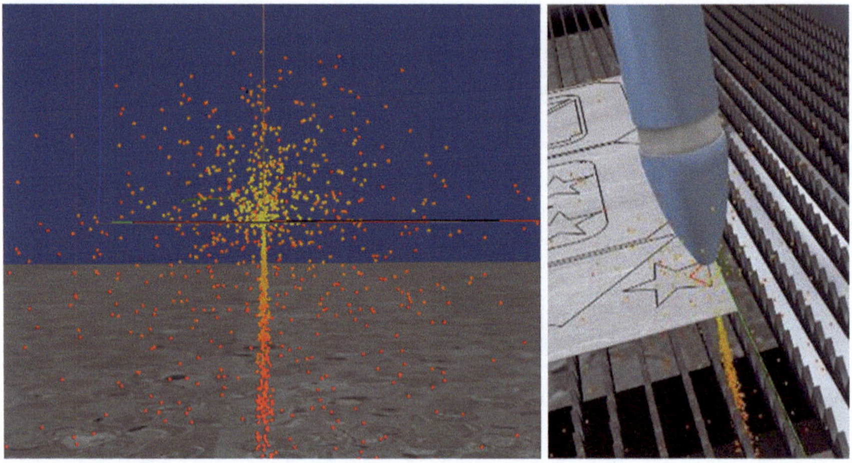

Fig. 5 Visual effect to represent the sparks in the machining processes. Sparks are produced by a GLSL shader, implementing a highly configurable particle system

3.4 Sparks Visual Effect

The machining process produces lot of sparks when they are cutting the metal sheet. Even in punching machines, some discrete sparks can be seen when the punch is performed.

The simulation system addresses this situation by adding a visual effect for the sparks. It is basically a combination of particle effects, developed as a GPU GLSL shader (Fig. 5). The visual effect can be configured to adapt the achieved result to different situations. For example, the colour, speed and spread angle of the particles can be adapted to match laser cutting sparks or water spread out from the torches.

4 Methodology

In the previous section the simulator features were described. The user interface and the 3D models of the machines were introduced. The missing element in the description was the definition of the metal sheet itself and how the machining operations are performed.

This section introduces the sheet representation and the cutting and punching operations support. Despite of being different processes, they are based on the removal of material from the sheet, so the internal module for such operations has been designed to be generic for such operations.

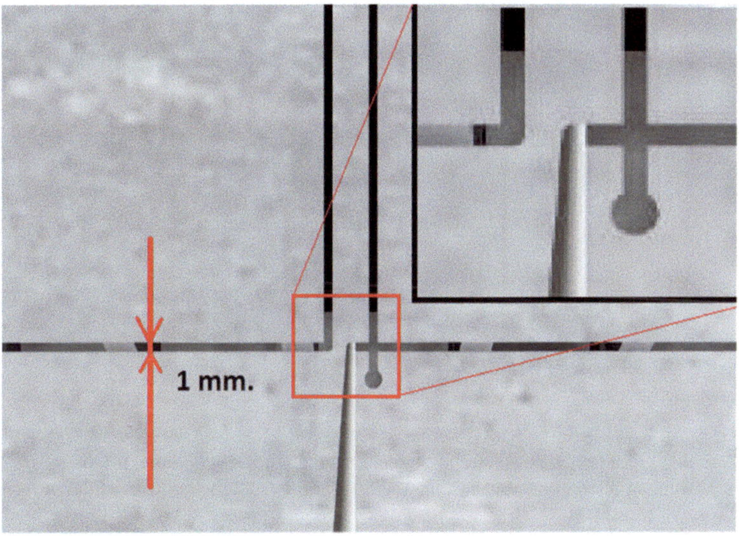

Fig. 6 Cutting and punching machining. In this case, the cutting process leaves a 1 mm. diameter path in the metal sheet

4.1 Sheet Representation

The sheet to be machined is conceptualized as a planar rectangle, defined by its contour. In fact, any given metal sheet can be defined as a set of non-intersecting and coplanar contours. Therefore, the machining operations will use 2D contour Boolean operation between the sheet object and the moving tools.

As a result, the high level 3D Boolean operation is simplified in a single 2D Boolean operation between two 2D complex polygons, which are a well-reviewed research topics [7, 10].

Normally, the metal sheet will start the machining process as a rectangular region, defined as a 4-vertices contour. As the process goes on, the path of the moving tool creates a sweep volume, which intersects the sheet. Although the torches cut the sheet in a very narrow path (0.5 mm or less), the swept area is noticeable (see Fig. 6).

The removed material from the sheet corresponds with the swept volume of the moving tool. Taking into account that we are dealing with 2D contours and that the torch is approximated as a very thin vertical cylinder, the moving tool can be generated directly for linear and arc movements (see Fig. 7).

The punching processes are essentially different, but the basic geometric operation remains the same (2D contour operations, see Fig. 8). The definition of the tool profile is given as a contour, as the tool shape is not a cylinder as in the cutting processes. The swept calculation is dropped in the punching machines, as the punches are discrete, i.e., the metal sheet does not move while the punching is being performed.

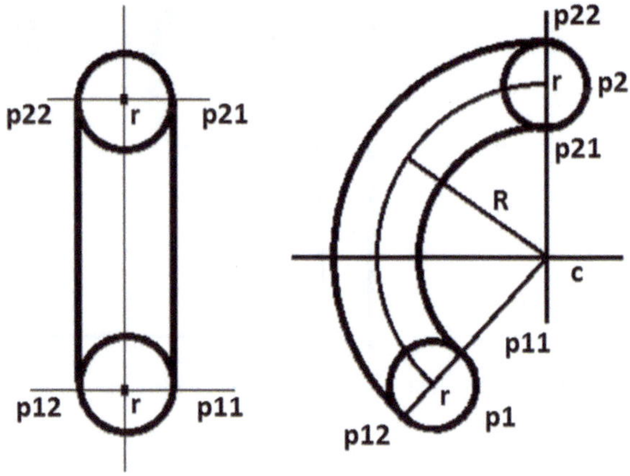

Fig. 7 A moving tool produces a sweep volume. Dealing with 2D contours, the swept area can be geometrically constructed given the starting and final points; and the radiuses of the movement

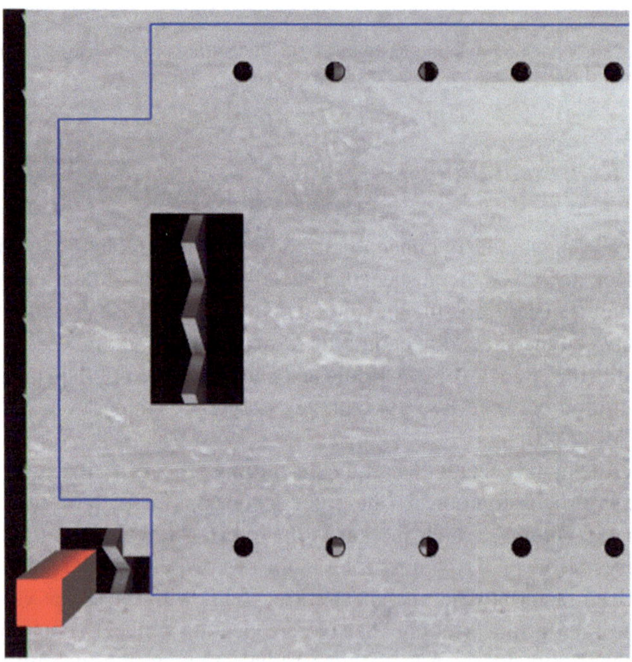

Fig. 8 A punching machine produces discrete holes in the metal sheet. The different tools have different shapes to allow complex products

Fig. 9 Render of 2D nested contours using OpenSceneGraph (GLUtesselator)

This situation simplifies the situation, as the calculation of a moving generic contour along a given path is not a straightforward task.

The Boolean operation between the current state of the metal sheet and the swept area by the tools is performed by the clipper library [4], a polygon clipping C++ software. The result of the subtraction is the next state of the metal sheet. This iterative process continues till the end of the machining instructions.

4.2 Rendering the Geometry

The visualization of the metal sheet relies on OpenSceneGraph, the used graphics engine utilised in the simulation system. In order to render the 2D contours into the screen, a tessellation of the 2D polygons, possibly with holes, is required. This process is internally handled by OpenGL *GLUtessellator* objects and methods.

Despite of the simplicity of the approach, the results show proper renderization of the polygons (see Fig. 9).

The metal sheets are very thin, but they have thickness. To render the 3D dimension of the metal sheet, a duplicated geometry is rendered (top cover) and perpendicular planes are created to join both geometries. This skinning process is just a matter of creating a *triangle_strip* geometry per contour.

4.3 Optimization Challenges

A direct implementation of the methods proposed in this section does not give great performance to the simulation system. As the simulation goes on, the metal sheet representation grows, in number of contours, and in number of vertices. Therefore, the Boolean operations become less efficient and the overall performance drops. Additionally, the rendering of very complex and nested contours is not an easy task for the underlying OpenGL subsystem, due to the tessellation process.

The next section will introduce a spatial partitioning system to improve the performance of the simulator.

5 Spatial Partition and Results

The efficiency of the polygon clipping algorithms depends directly on the total number of contours and points involved in the Boolean operation [5]:

$$O(n \times \log(n) + k + z \times \log(n))$$

where n is the number of edges (points), z is the number of contours and k is the number of edge intersections. As the simulation is performed, the working part gets more and more complex and consequently, the number of points and contours grows. In order to limit the number of points and contours that would increase the Boolean operation time, a high-level partitioning system is used.

This spatial partitioning decomposes the metal sheet into a set of smaller subregions, leading to a high-level Boolean operation pseudo-algorithm. The performance effect of the spatial subdivision is limited, as an over-subdivided sheet will increase the number of individual Boolean operations, as any movement will span across multiple regions. The subdivision region is aimed to reduce the complexity of the Boolean operation (by limiting the number of vertices) but avoiding increasing the mean number of Boolean operations per movement.

In our experiments (see Table 1), varying the number of subdivision with the same example, gives a performance peak using the 16×16 subdivision. The Fig. 10 shows the geometric complexity of the metal sheet after the simulation program is completely executed. Approximately, each full circle contains around 100 vertices, 50 for the internal circle and 50 for the external circle. The whole geometric sheet contains around 1 million vertices, and due to the spatial subdivision (16×16 in this case), the rendering is performed in real time using the OpenSceneGraph's internal polygon tesselator (OpenGL's GLUtesselator methods).

Table 1 Spatial subdivision performance, with the direct sweep arc generation applied, varying the spatial subdivision

Model	Movements	Subdivision	OpBools	Time (s)
Simple	834	1 × 1	774	10
		4 × 4	1,008	5
		16 × 16	2,700	6
		32 × 32	6,556	10
Complex	12,731	1 × 1	11,037	2,100
		4 × 4	12,481	195
		16 × 16	19,533	87
		32 × 32	36,910	130

The number of low level boolean operation and the time are presented in the columns OpBools and time

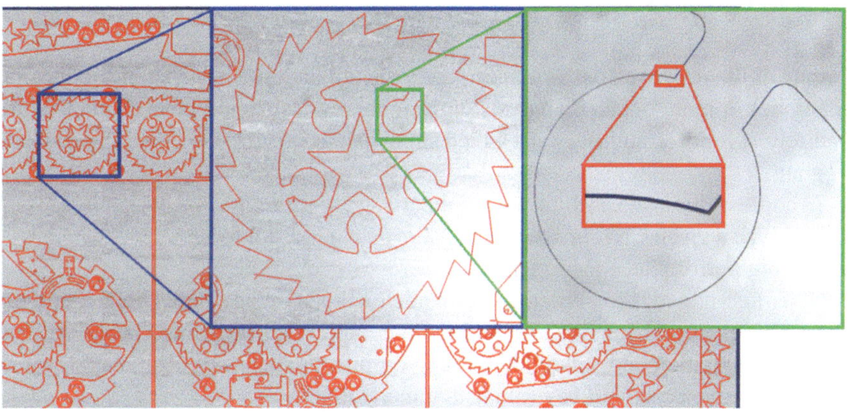

Fig. 10 Simulation result of a large CNC program (12,000 movements)

6 Conclusions

In this work we have introduced a representation based on 2D complex polygons to represent the metal sheet in machining processes. All the operations of the programs have been transformed to internal Boolean operations between the sheet and the sweep of moving cylinder (cutting machines) or a complex polygon (for punching machines).

As the efficiency of the Boolean operations decay with the number of points and polygons, we have introduced several mechanisms to optimize the overall result of the simulation. To limit the number of points and polygons, a spatial partition system has been used, checking that it cannot be increased to arbitrary numbers, since it will provoke a huge explosion in the number of Boolean operations. A balanced spatial partition is required to get the best performance.

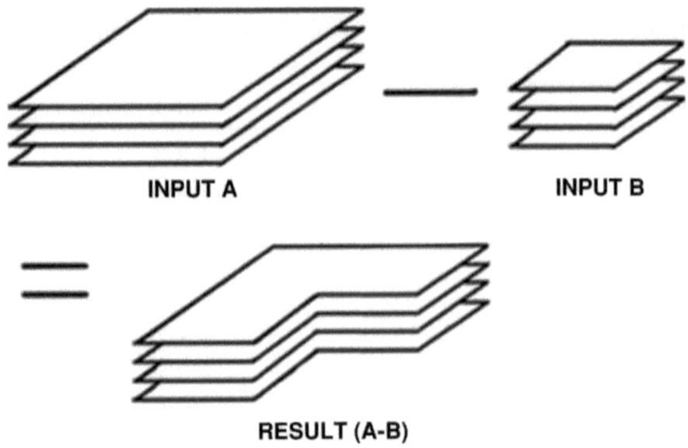

Fig. 11 A Boolean operation example between two 3D objects represented as a set of parallel levels

7 Future Works

Despite of the improved performance due to the utilisation of the spatial subdivision techniques, a more detailed analysis of such conditions should be evaluated. More precisely, it is not straightforward which subdivision is the best one for a given CNC program.

To address this issue, as future work, heuristic algorithm will be researched. They will analyze a given CNC program and estimate a good subdivision level which will bring a big performance increment.

Other alternative is to implement dynamic methods in the subdivision technique. As a given region becomes more and more complex, a background process could detect that situation and subdivide that region into smaller ones. The overall result in the performance should be evaluated numerically.

An extension of the proposed methodology to 3D could be addressed. A 3D object can be approximated as a set of parallel levels (see Fig. 11). Each level would contain a set of 2D contours, as the metal sheet used in this work. This extension will bring new problems into the field, such as, skinning algorithms to create a proper visualization of the 3D object, proper sweep volume calculations and conversion to a set of 2D contours and impact in the performance.

Acknowledgments We thank the Basque Government Industry Department for the financial help received under the GAITEK research programme.

References

1. Brunet, P., Navazo, I.: Solid representation and operation using extended octrees. ACM Trans. Graph. **9**(2), 170–197 (1990)
2. Cano, P.: Representation of polyhedral objects using sp-octrees. J. WSCG **10**(1), 95–101 (2002)
3. Hook, V.: Real time shaded NC milling display. SIGGraph86 20(4), 15–20 (1986).

4. Johnson, A.: Clipper–an open source freeware polygon clipping library. http://www.angusj.com/delphi/clipper.php (2012)
5. Leonov, M.V.: Implementation of boolean operations on sets of polygons in the plane (1998).
6. OpenSceneGraph.: Open source 3D graphics API over OpenGL. http://www.openscenegraph.org/ (2012)
7. Preparata, F.P., Shamos, M.I.: Geometry: An Introduction. Springer, Berlin (1985). ISBN 0-3879-6131-3
8. Spence, A.D., Li, Z.: Parallel processing for 2–1/2D machining simulation. In: Proceedings of the Sixth ACM Symposium on Solid Modeling and Applications. ACM, SMA '01. pp. 140–148, ISBN 1-58113-366-9 (2001).
9. Stewart, N., Leach, G., John, S.: Improved CSG rendering using overlap graph subtraction sequences. In: International Conference on Computer Graphics and Interactive Techniques in Australasia and South, East Asia, pp. 47–53 (2003).
10. Vatti, B.R.: A generic solution to polygon clipping. Commun. ACM **35**(7), 56–63 (1992)
11. Wallace, E.: Constructive solid geometry on meshes using BSP trees in JavaScript. http://evanw.github.com/csg.js (2012)
12. Zhu, W., Lee, Y.: Product prototyping and manufacturing planning with 5-DOF haptic sculpting and dexel volume updating. In: Haptic Interfaces for Virtual Environment and Teleoperator Systems, pp. 98–105 (2004).

The Case for Scripted Process Design and Engineering

Thomas A. Grandine

Abstract This paper summarizes recent research published by Tim Simpson and his colleagues. The original intent of the research was to quantify the benefits of interactive design tools. For various reasons, the research was unsuccessful in accomplishing this objective, possibly for the simple reason that the benefits may not be as real as previously believed. This paper concludes with advocacy for another well-established idea, namely design through careful development of scripted design processes, frequently referred to as Knowledge Based Engineering throughout the literature.

Keywords Geometric design · Knowledge-based engineering · Wing planform · Wing optimization · Script-based design

1 Introduction

Interactive design tools first became popular in the 1970s, and their use has continued to expand ever since. Today, they have become ubiquitous in engineering design, and they are assumed by nearly everyone to be the most efficient and effective means of creating geometry models for product definition, engineering analysis, and automated and semi-automated manufacturing. Their use is taught in most engineering departments and schools, and they are considered the best tools currently available for the job of creating not only singular designs, but also parametric families of designs for design optimization trade studies and multi-member product families [2].

Nevertheless, little hard evidence exists to back up the widely held belief that they are the right tool for many of the geometric model building tasks for which they are used. Indeed, some research attempting to generate this evidence has been performed, most notably by Tim Simpson and his students in a series of papers

T. A. Grandine (✉)
The Boeing Company, P. O. Box 3707, MC 42-58, Seattle, WA 98124, USA
e-mail: thomas.a.grandine@boeing.com

R. De Amicis and G. Conti (eds.), *Future Vision and Trends on Shapes, Geometry and Algebra*, Springer Proceedings in Mathematics & Statistics 84,
DOI: 10.1007/978-1-4471-6461-6_10, © Springer-Verlag London 2014

[1, 4–8]. This careful research, performed with live subjects under carefully controlled circumstances is remarkable in that it produced essentially no evidence bolstering the claim that interactive design tools have increased productivity and design quality. Indeed, as I look over the research, I can see hints that the opposite might very well be true, though the authors of this careful work are careful not to engage in that particular speculation.

This paper is divided into two main sections. Section 2 reviews the findings from one of the studies performed by Simpson and his colleagues at Boeing, using real Boeing engineers and mathematicians. Section 3 explores the question of why the Simpson studies failed to quantify the benefits of interactive design tools, and it resurrects the well-established alternative idea of script-based design tools and explains why that idea may prove, in the long run, to be better suited to engineering design than today's interactive tools. The discussion is concluded in the Sect. 4.

2 Simpson's Findings

Simpson's early studies quantifying the benefits of interactive design tools focused on three small design problems: An I-beam, a desk lamp, and the layout of a shop floor. All three design problems involved only two or three design variables so that visualization of the entire design space was feasible. The researchers developed interactive applications for each problem, complete with a graphical user interface, that enabled the users to adjust the values of the design variables with a slider bar and visualize not only the resulting design, but also important engineering analysis results. Equipped with this software, the researchers found student test subjects who, after a nominal amount of training, were asked to use the interactive tool to determine a good solution to the given problem. After each subject finished his or her design, he was given a questionaire to complete. Results were then compiled by the researchers based on how well the subjects had performed and their answers to the survey questions. See [1, 4, 5, 7] for additional details.

In each case, test subjects were given one of three different versions of the interactive design tool. One version provided immediate visual feedback whenever a design variable slider bar was moved. Both the geometry and the analysis results were updated instantaneously. In a second version of the software, test subjects experienced a slight delay in refresh of the graphics, while in a third version, the delay was more substantial. While the test subjects reported increasing levels of frustration as the length of the delay increased, the researchers found no correlation between this delay and the quality of the designs they produced. In other words, zippy interactive design tools appeared to have an effect on job satisfaction, but not on the quality of the work performed.

The results surprised the researchers. As modest individuals, they attributed their frustrating lack of evidence on the benefits of interactive design to lack of care and control in the experiments they performed. They seem not to have considered the provocative idea that their lack of evidence is due to the fact that the hypothesis they

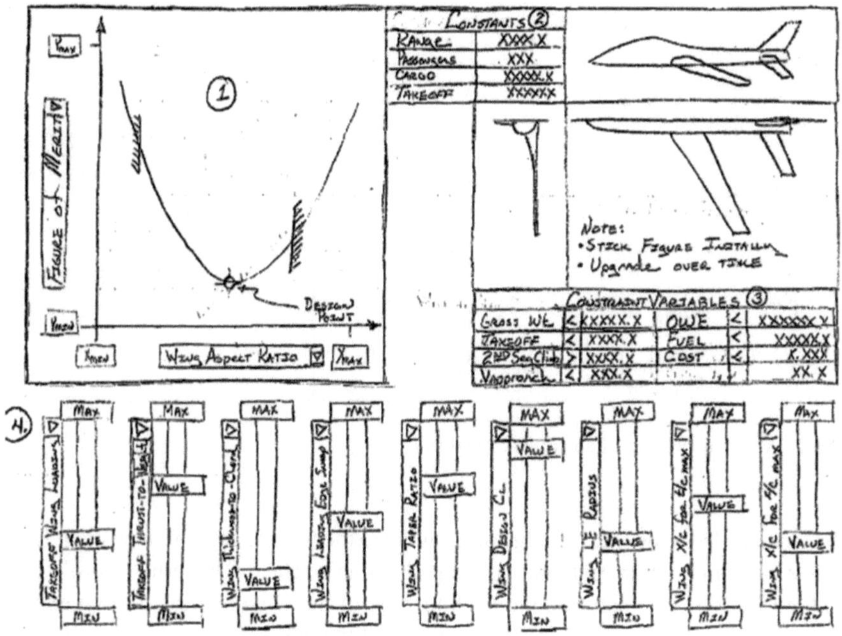

Fig. 1 A sketch of the desired user interface

set out to prove simply isn't true. Consequently, they designed additional experiments which controlled additional aspects of the applications and user experience in an attempt to tease out some of the quantitative information they were seeking. One idea was to perform an experiment using experienced engineers rather than graduate students, so Simpson collaborated with Martin Meckesheimer, then a Boeing statistician, to conduct an experiment at Boeing.

The problem selected was a six variable wing design problem for a commercial jet transport configuration. One Boeing engineer with considerable design experience sketched his idea for what he thought a useful and attractive graphical interface for such a problem might look like. That sketch is shown in Fig. 1.

The design problem is to optimize the shape of an airplane wing planform. The planform of a wing is the 2D projection of the wing onto the wing reference plane, a nearly horizontal plane rotated a few degrees about the roll axis of the airplane. The purpose of this slight adjustment of a few degrees is to enhance the airplane's roll stability. The six variables in this particular airplane trade study are wing semispan, aspect ratio, quarter chord sweep angle, taper ratio, spar box root chord, and fan diameter. The wingspan of an airplane is the distance from wingtip to wingtip, and wing semispan is half that distance, or the distance from the centerline of the airplane to one of the wingtips. The aspect ratio is defined to be the wingspan squared divided by the area of the wing planform; it is essentially a measure of how long and narrow the wing is. The quarter chord sweep angle is the angle of the line which is 25 % of

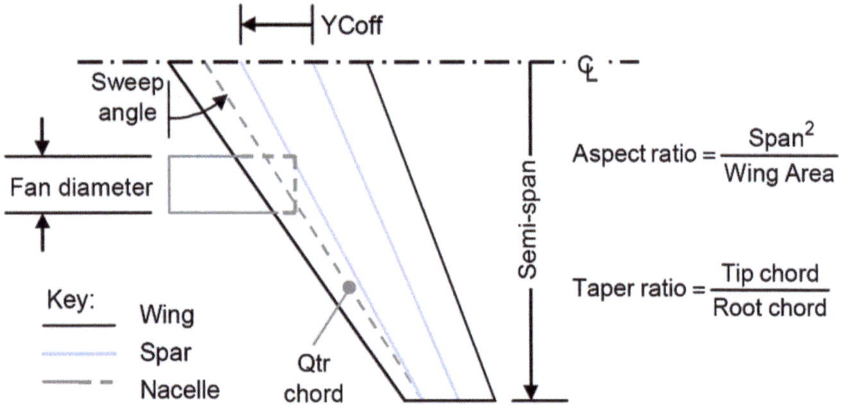

Fig. 2 A schematic view of an airplane planform

the distance from the leading edge of the wing to the trailing edge of the wing relative to a wing which is perpendicular to the body of an airplane. The term "chord" refers to the local distance between the leading edge of the wing and the trailing edge of the wing at any given distance along the span of the wing. The taper ratio is the ratio of the chord at the wingtip to the chord at the root of the wing, i.e. at the centerline of the airplane body. Wings are often supported by a pair of spars which run the length of the wing at some distance aft of the leading edge and forward of the trailing edge. The spar box chord is the distance between the forward spar and the aft spar. In this problem, the spar box root chord is one of the design variables. Finally, the diameter of the engine fan is a design variable. The setup for this problem is depicted in Fig. 2.

The interactive application contains a means of evaluating four analysis functions which depend on the six design variables. These are cost, an estimate of the total cost to build an airplane, range, an estimate of how far any given candidate design can fly without refueling, buffet altitude, the altitude at which airplane maneuvering begins to be difficult, and takeoff field length, an estimate of how much runway is needed for successful takeoff under various adverse conditions. The (non-dimensionalized) formulation of the optimal design problem is given by

$$\begin{aligned} \text{minimize}\ & \text{cost} \\ \text{subject to}\ & \text{range} > 0.589 \\ & \text{buffet altitude} > 0.603 \\ & \text{takeoff field length} < 0.377 \end{aligned}$$

As in the earlier studies, Simpson and his colleagues constructed an interactive program to aid solution for this design problem. A screen capture of this application is shown in Fig. 3.

In this application, note that there are six slider bars for each of the six design variables as well as a text entry box for each, enabling specific values to be entered directly. Above the slider bars is a display of the actual geometry of the current

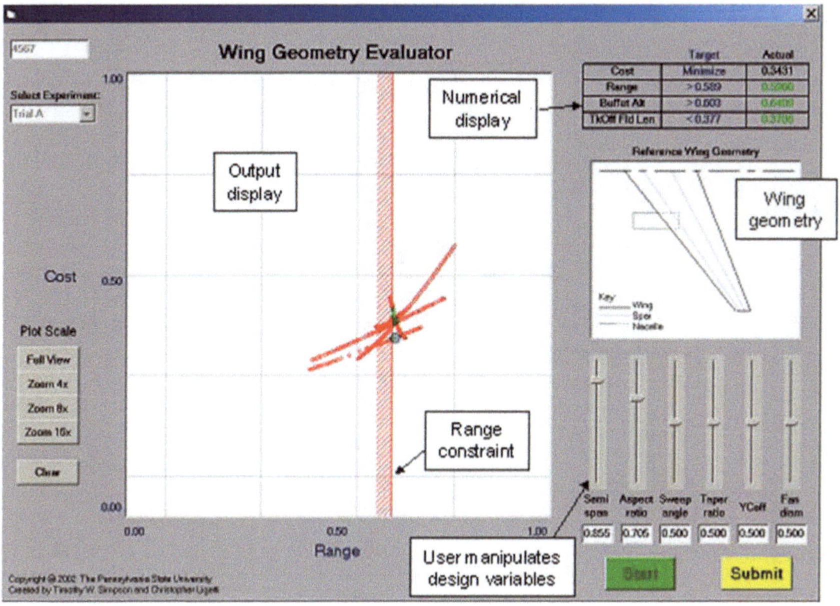

Fig. 3 An interactive application for designing an airplane planform

design, while the main display depicts a plot of the airplane range versus its cost. Moving the slider bar causes additional points to be plotted in the main window. The objective is to find the lowest point which is on or to the right of the vertical line which depicts the range constraint. The other two constraints are shown textually directly above the plot of the airplane geometry. The computed values for the three constraints are shown in green for constraints that are satisfied, whereas violated constraints have their values shown in red.

As in earlier studies, Simpson's team provided three versions of the software featuring three different levels of responsiveness. One version featured real-time updating of the display, and the other two versions featured 1/4 and 1/2 s delays, respectively. Each of the 22 participants in the study was assigned one of the three versions of the software and was given a few minutes of training on its use. Following the training, users were asked to spend as much time with the application as needed in order to design the best airplane, then press the submit button. After submitting their designs, users were given a questionaire about their experience, and their opinions, along with their performance data on the study, were studied statistically.

To the surprise of the team, the study was every bit as inconclusive as the previous, simpler studies. Indeed, the results may even have been more discouraging in that slight hints that interactive design was counterproductive were in evidence. For one thing, delay had no significant effect on error. In fact, the group which submitted the best designs overall was the group with the largest interactive response delay, though that better average performance was not statistically significant to the other groups.

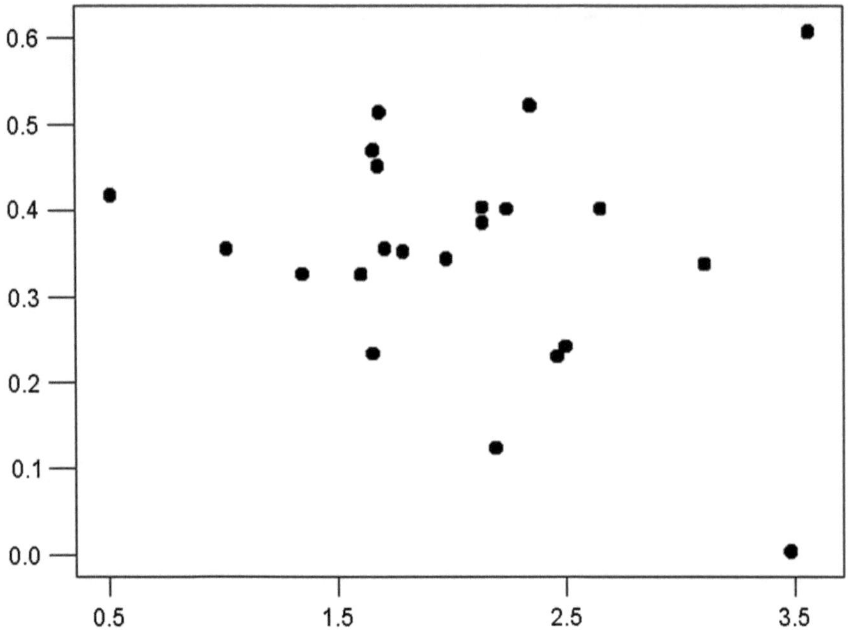

Fig. 4 Performance of the Boeing test subjects plotted against time taken to complete the exercise

Moreover, the interactive delay of the tool had no meaningful effect on completion time, with the group with the 1/2 s delay finishing almost as fast as the group with no delay. The group with the 1/4 s delay was the slowest and poorest performing group overall on average.

The researchers found that there was no correlation between time spent with the tool and the quality of the design produced. These results are shown in Fig. 4. The horizontal axis represents non-dimensionalized time to complete the exercise, while the vertical axis represents the discrepancy between the best possible design and the one chosen by the participant. This chart shows the 22 participants scattered all over. Indeed the very best performer and the very worst performer in terms of quality of design happened to be the two individuals who spent the most time completing the exercise.

The researchers were able to assert three things with some confidence: (1) There was a strong correlation between familiarity with use of slider bars and how well individuals did, (2) familiarity with aircraft design and layout correlated with how well individuals performed on the assessment, and (3) how frequently individuals made use of the table of values which contained the computed objective function and constraint function values and the number of designs visited. Seven additional correlations which the researchers noted but had somewhat less confidence in were also reported. Two of these are anecdotally interesting. There was an inverse correlation between how confident individuals were in the quality of their designs and the actual quality, with those individuals expressing less confidence substantially outperforming those individuals who thought they did well. In addition, there was a correlation between the individuals who frequently made use of the picture of the actual airplane geometry and the length of time taken to complete the exercise. Individuals using that picture took longer than the individuals who did not, but there was no correlation between use of that picture and quality of design submitted.

The researchers made one additional observation: "We found that the top five performers, in terms of lowest error, were non-engineers .... It became evident during testing that all of these users exploited the interactions that existed between the independent design variables in order to find better designs; few, if any, of the engineers took advantage of these interactions, leading to worse designs." In hindsight, this is perhaps not so surprising. After all, Simpson and his colleagues had already done the hard engineering work of formulating the design problem as an optimization problem. Once the design problem had been reduced to a math problem the surprise would have been had the mathematicians not outperformed the engineers on the study.

3 Observations

Why were the researchers unsuccessful in quantifying the benefits of interactive design over the course of several experiments? Their own belief is that their experiments were not designed carefully enough and that they weren't measuring the right things. There is surely some truth to this since, as observed in the preceeding paragraph, the experiments really measured the effectiveness of solving a well-posed mathematics problem interactively rather than an open-ended engineering problem. Unsurprisingly, moving slider bars proves not to be an effective means of solving well-posed mathematics problems which are more amenable to algorithmic techniques.

I doubt that this entirely accounts for their lack of success, however. Specifically, I suspect that the allure of interactive design tools is due more to their apparent ease of use than to cost effectiveness. This suspicion is borne out by the findings of Simpson and his colleagues, too, in that on every survey, those assigned tools without any interactive delay response expressed greater satisfaction with the design tool than those who were given the versions of the tool with noticeable delays. That this satisfaction did not translate into better design results suggests that ease of use and interactivity do not correlate with effectiveness, a possibility the researchers were never willing to confront.

This possibility seems very real to me. Consider the three airplane fuselages designed interactively and shown in Fig. 5. The black lines show how the underlying mathematical models are pieced together to form a complete fuselage model. Aside from the fact that all of the models have a large number of individual pieces, no discernible consistent strategy has been applied to their construction. All were built using interactive tools for which the most important measure of success was the

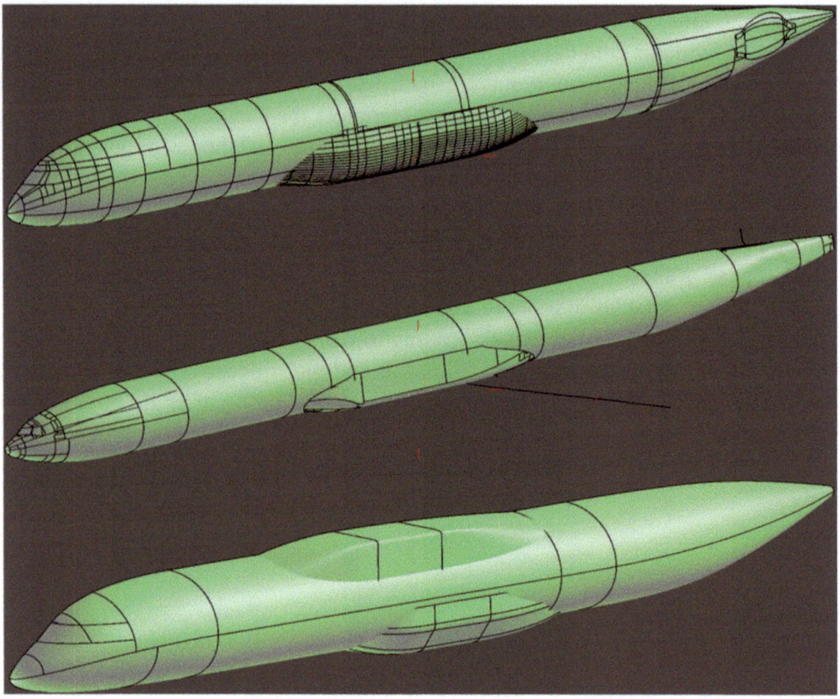

Fig. 5 Three different interactively constructed airplane fuselage models

quality of the image that appeared on the screen (without the black lines, of course). Because the usability of the model by the downstream applications, whether they be manufacturing applications, engineering analysis applications, financial applications, or customer presentation cannot be assessed visually, it typically doesn't get performed by the operator of an interactive tool.

If interactive design methods aren't the productivity boon they are widely believed to be, then what is the right strategy? My own experience suggests that careful development of scripted processes which automate important design tasks is a much better idea. This is not a new idea, and it has gone by many different names over the years, perhaps the most important of which is Knowledged-Based Engineering (KBE). Several commercial systems are available which implement this technology in varying degrees, and there is a considerable literature associated with it. The paper [3] gives a great overview of this topic in an important application area.

Although experts in the area have disagreements about exactly what KBE encompasses, all agree that it surely includes the writing of programs in high-level languages which capture important design process knowledge so that it can be reused in the future. The title of this paper uses the phrase "scripted process design" to capture this important idea that is essential to KBE.

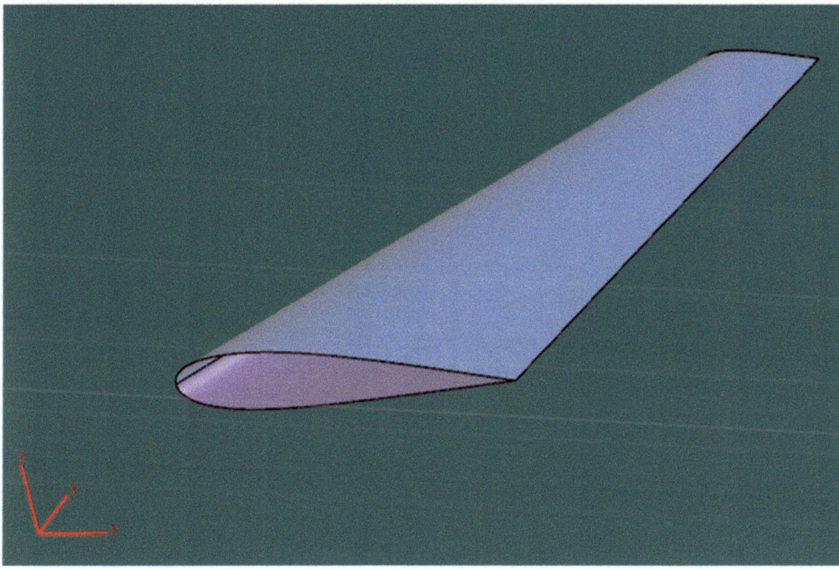

Fig. 6 *Upper* and *lower* surfaces which form an airplane wing

There is no silver bullet here. The very best tools for capturing process knowledge of various kinds are computer programming languages. Indeed, that is the very thing that they are designed to do, and no better means of capturing process knowledge has ever been devised. Powerful and effective point and click tools for writing computer programs simply do not exist, so it is not surprising that capturing design process information interactively through graphical users interfaces has proved elusive. The good news is that improvements and refinements in computer programming continue to be made, and modern languages offer many features and facilities that simultaneously increase capability and ease of use.

Boeing has a home grown geometry scripting capability that is based on the Python programming language. As an example of the power of this idea, consider the following problem: Imagine a pair of surfaces s_1 and s_2 that together form a wing-like enclosure (see Fig. 6). Suppose that s_1 is the upper part of this enclosure, while s_2 is the lower part. Further suppose that points on the surfaces are given by $s_1(u_1, v_1)$ and $s_2(u_2, v_2)$. Each of the surfaces has component functions, so

$$s_1(u_1, v_1) = \begin{pmatrix} x_1(u_1, v_1) \\ y_1(u_1, v_1) \\ z_1(u_1, v_1) \end{pmatrix}.$$

Finally, suppose that both surfaces are parametrized in such a way that the second parameter increases with the span of the wing, i.e. $y_1(u_1, v_1) = y_2(u_2, v_2)$ whenever $v_1 = v_2$. For each value of v_1, the vertical distance between the upper and lower surfaces will be maximized whenever the corresponding tangent vectors are parallel, i.e. when

$$\frac{\partial}{\partial u_1}\begin{pmatrix} x_1(u_1, v_1) \\ z_1(u_1, v_1) \end{pmatrix} \parallel \frac{\partial}{\partial u_2}\begin{pmatrix} x_2(u_2, v_2) \\ z_2(u_2, v_2) \end{pmatrix}.$$

With this in mind, new surfaces $\widehat{s_1}$ and $\widehat{s_2}$ can be defined

$$\widehat{s_1}(u_1, v_1) = \begin{pmatrix} x_1(u_1, v_1) \\ y_1(u_1, v_1) \\ \dfrac{\frac{\partial}{\partial u_1} z_1(u_1, v_1)}{\frac{\partial}{\partial u_1} x_1(u_1, v_1)} \end{pmatrix}$$

$$\widehat{s_2}(u_2, v_2) = \begin{pmatrix} x_2(u_2, v_2) \\ y_2(u_2, v_2) \\ \dfrac{\frac{\partial}{\partial u_2} z_2(u_2, v_2)}{\frac{\partial}{\partial u_2} x_2(u_2, v_2)} \end{pmatrix}.$$

Note that the third components of each of these new surfaces is just the slope of the curves which are the sections of the surfaces. The tangent vectors will be parallel exactly when these slopes are the same. Thus, the desired curves along each of the two original surfaces s_1 and s_2 can be calculated as the intersection of the new surfaces $\widehat{s_1}$ and $\widehat{s_2}$. Here is a short Python code written in Boeing's geometric scripting language which accomplishes this

```
## Define function to create modified surfaces

def hatsurf (surf):
    surf = surf.Trim ([[0.05, 0.95], [0, 1]])
    surfu = surf.Differentiate ()
    xy = surf.Multiply ([[1, 0, 0], [0, 1, 0], [0, 0, 0]])
    xu = surfu.Multiply ([[1, 0, 0]])
    zu = surfu.Multiply ([[0, 0, 1]])
    newz = (zu / xu).Multiply ([[0], [0], [1]])
    return xy + newz

## Create hat surfaces and intersect

s1hat = hatsurf (s1)
s1hat = hatsurf (s2)
([s1trace, s2trace]) = Intersect (s1hat, s2hat)
```

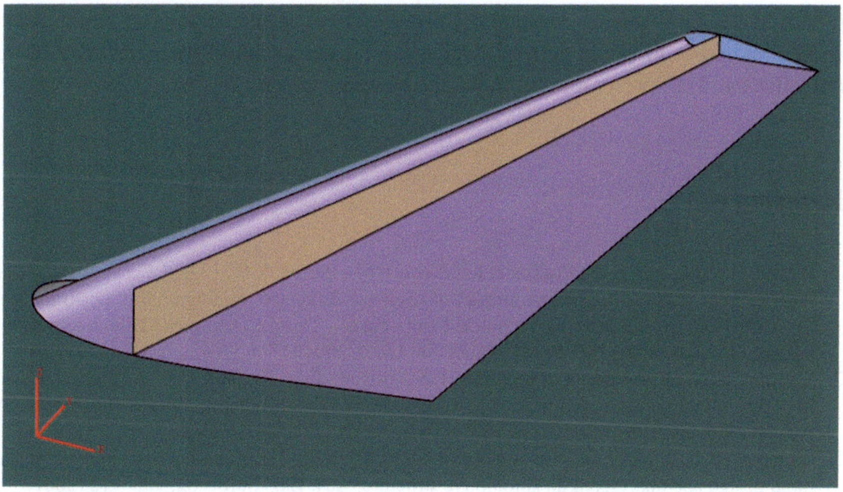

Fig. 7 Wing showing maximum depth spar. The *upper* wing surface is not shown so that the spar is visible

```
## Create new spar surface

s1curve = s1.Compose (s1trace.GetMap ())
s2curve = s2.Compose (s2trace.GetMap ())
spar = RuledSurface (s1curve, s2curve)
```

The result of this script is the surface shown in Fig. 7. This script is designed to work for all wings built according to the specifications mentioned earlier. Since it is independent of the actual geometry of the airplane wings, it provides an automated means of producing a spar whenever the geometry of one or both of the wing surfaces changes.

4 Conclusion

The combination of Simpson's studies and my own years of experience watching engineers use interactive design tools makes clear that interactive tools do not (1) reduce design time, (2) lead to better designs, (3) empower engineers, (4) enhance repeatability of the design process, (5) support downstream engineering and manufacturing applications, (6) capture engineering knowledge, or (7) institutionalize preferred design practices. Despite requiring more expertise on the part of the users and being harder to use in general, script-based tools accomplish all seven of these objectives and are preferable for most design applications.

References

1. Barron, K., Simpson, T.W., Rothrock, L.., Frecker, M., Barton, R.R., Ligetti, C.: Graphical user interfaces for engineering design: Impact of response delay and training on user performance. In: ASME Design Engineering Technical Conferences—Design Theory & Methodology Conference, Salt Lake City, UT. ASME, Paper No. DETC2004/DTM-57085 (2004)
2. Bowcutt, K.G., Kuruvila, G., Grandine, T.A., Hogan, T.A., Cramer, E.J.: Advancements in multidisciplinary design optimization to hypersonic vehicles to achieve closure. In: 46th AIAA Aerospace Sciences Meeting, Reno, NV. AIAA, AIAA-2008-2591 (2008)
3. Chapman, C.B., Pinfold, M.: The application of a knowledge based engineering approach to the rapid design and analysis of an automotive structure. Adv. Eng. Softw. **32**, 903–912 (2001)
4. Frecker, M., Simpson, T.W., Goldberg, J.H., Barton, R.R., Holewinski, B., Stump, G.: Integrating design research into the classroom: Experiments in two graduate courses. In: 2001 Annual ASEE Conference, Albuquerque, NM. ASEE (2001)
5. Ligetti, C., Simpson, T.W.: Metamodel-driven design optimization using integrative graphical design interfaces: results from a job shop manufacturing simulation experiment. ASME J. Comput. Inf. Sci. Eng. **5**(1), 8–17 (2005)
6. Ligetti, C., Simpson, T.W., Frecker, M., Barton, R.R., Stump, G.: Assessing the impact of graphical design interfaces on design efficiency and effectiveness. ASME J. Comput. Inf. Sci. Eng. **3**(2), 144–154 (2003)
7. Simpson, T.W., Peplinski, J., Koch, P.N., Allen, J.K.: Metamodels for computer-based engineering design: survey and recommendations. Eng. Comput. **17**(2), 129–150 (2001)
8. Simpson, T.W., Meckesheimer, M.: Evaluation of a graphical design interface for design space visualization. In: 45th AIAA/ASME/ASCE/AHS/ASC Structures, Structural Dynamics & Materials Conference, Palm Springs, CA. AIAA, AIAA-2004-1683 (2004)

MIX
Papier aus verantwortungsvollen Quellen
Paper from responsible sources
FSC® C105338

If you have any concerns about our products,
you can contact us on
ProductSafety@springernature.com

In case Publisher is established outside the EU,
the EU authorized representative is:
Springer Nature Customer Service Center GmbH
Europaplatz 3, 69115 Heidelberg, Germany

Printed by Libri Plureos GmbH
in Hamburg, Germany